Archaeology Data Service

GIS *Guide to Good Practice*

Archaeology Data Service

GIS *Guide to Good Practice*

edited by Mark Gillings and Alicia Wise

with contributions by Mark Gillings, Peter Halls,
Gary Lock, Paul Miller, Greg Phillips, Nick Ryan,
David Wheatley and Alicia Wise

Produced by Oxbow Books for the Arts and Humanities Data Service

ISBN 1 900188 69 4
ISSN 1463–5194

This book is available direct from

Oxbow Books, Park End Place, Oxford OX1 1HN
(Phone: 01865–241249; Fax: 01865–794449)

and

The David Brown Book Company
PO Box 511, Oakville, CT 06779, USA
(Phone: 860–945–9329; Fax: 860–945–9468)

or from our website
www.oxbowbooks.com

Printed in Great Britain at
The Short Run Press
Exeter

Contents

Acknowledgements

We would like to express our sincere thanks to all the people who reviewed the GIS Guide to Good Practice, offered suggestions for its improvement, and contributed to its shape.

Mark Aldenderfer, University of California – Santa Barbara

Javier Baena, Universitaria de Cantoblanco

Bob Bewley, Royal Commission on the Historical Monuments of England

Nick Burton, Central Archaeology Service, English Heritage

Steve Catney, Association of Local Government Archaeological Officers

Nigel Clubb, Royal Commission on the Historical Monuments of England

David Dawson, Museums Documentation Association

Piers Dixon, Royal Commission on the Ancient and Historical Monuments of Scotland

Danny Donaghue, University of Durham

Vince Gaffney, University of Birmingham

Mike Heyworth, Council for British Archaeology

Jeremy Huggett, University of Glasgow

Val Kinsler, 100% Proof

Edmund Lee, Data Standards Unit, Royal Commission on the Historical Monuments of England

Ian Morrison, English Heritage

Ian Oxley, Archaeological Diving Unit, University of St Andrews

Gillian Quine, Data Standards Unit, Royal Commission on the Historical Monuments of England

Les Rackham, Ordnance Survey

Julian Richards, University of York

Armin Schmidt, University of Bradford

Tim Seaman, Archeological Records Management Section, New Mexico Historic Preservation Division

Zoran Stancic, University of Ljubljana

Martijn van Leusen, University of Groningen

Rob Walker, Association for Geographic Information

Section 1: Aims and Objectives

1.1 WHY A GIS *GUIDE TO GOOD PRACTICE*?

This document is designed specifically to provide guidance for individuals and organisations involved in the creation, maintenance, use and long-term preservation of GIS-based digital resources. It should be noted that although the overall emphasis is upon archaeological data, the information presented has much wider disciplinary implications.

As well as providing a source of useful generic information, the guide emphasises the processes of long-term preservation, archiving and effective data re-use. As a result, the importance of adhering to recognised standards and the recording of essential pieces of information about a given resource are dominant recurring themes throughout the discussions. The latter are designed to smooth the transition of the digital resource into an archive environment, in particular the Archaeology Data Service (ADS), and to ensure that it can be re-located and re-used in the future.

It is important to realise that the current document is one of a family of theme-specific guides, some of which contain much more detailed discussions of many of the topics outlined here, such as the integration of satellite images, and the precise formats and convention standards used within CAD layers. Although each guide, whether concerned with GIS, CAD, geophysical survey or remote sensing, is specifically designed to be self-contained, taken together they comprise a comprehensive, authoritative and highly complementary set of practical guidelines.

In outlining the aims of the document, it is equally important to state what the current guide does not cover. It does not aim to constitute an exhaustive introduction to the underlying origin, theory and technical implementation of GIS. Nor is it in any way a definitive and prescriptive manual on how 'best' to undertake GIS. Although the importance of standards and data frameworks will be rightly emphasised, the aim of the guide is to introduce practitioners to areas and issues where standards and frameworks already exist and may be applicable, and to identify the relevant sources of information that may be consulted. Whilst optimum pathways will often be identified, the guide does not rigidly advocate any single standard or narrow set of options. Instead, the concern here is more generic, with the aim of encouraging and developing the routine use of standards and data frameworks as a whole. In this sense it is important to realise that the present document constitutes a *guide* as opposed to a *manual*.

It is also important to note that this guide is concerned solely with archaeological data and GIS, whether derived from excavation, regional survey, archival research, intra-site analysis or any other archaeological endeavour. It is not concerned with the integration, archiving and accessing of data destined for study, maintenance and future re-use within CAD systems. This topic will be covered in detail in the forthcoming CAD and Excavation and Fieldwork guides to good practice.

This *Guide to Good Practice* is one of five commissioned by the Archaeology Data Service. The five archaeological guides consist of a general guide to Digital Excavation and Fieldwork Archiving guide as well as four more specific guides. These are on Aerial Photography and Remote Sensing, CAD, and Archaeological Geophysics, as well as this volume on GIS.

The ADS is a part of the Arts and Humanities Data Service (AHDS), which caters for digital archiving needs across the humanities disciplines of archaeology, history, literary studies, performing arts, and visual arts. Each branch of the AHDS takes responsibility for advising on good practice in the creation, management, preservation, and re-use of digital information widely used in its subject area. Planned titles (across the AHDS) include the following guides:

Service Provider	Title of *Guide*
Archaeology Data Service	Archaeological geophysics: a guide to good practice
	Archiving aerial photography and remote sensing data: a guide to good practice
	CAD guide to good practice
	Digital archives from excavation and fieldwork: a guide to good practice
	GIS guide to good practice
History Data Service	Digitising history: a guide to creating digital resources from historical documents
	Mapping the past: a basic guide to GIS for historians
Oxford Text Archive	Creating and documenting electronic texts
	Developing linguistic corpora
	Finding and using electronic texts
Performing Arts Data Service	Guide to good practice in creating digital audio resources
	Guide to good practice in creating digital performance resources
Visual Arts Data Service	Creating digital resources for the visual arts: standards and good practice
	Using digital information in teaching and learning in the visual arts
Cross-Disciplinary Guide (led by VADS)	Creating and using virtual reality: a guide for the arts and humanities

1.2 HOW BEST TO USE THIS GUIDE

Ideally any individual or institution involved with, or planning, a GIS-based exercise with the long-term aim of depositing the resultant data with the ADS should read the guide in its entirety. In many cases, however, practitioners will only be dealing with one particular stage of the overall process, and to reflect this the guide has been structured into clear thematic sections. Individuals are advised to read the sections relevant to the task at hand carefully.

As has been mentioned, throughout the sections are lists of information that it is critical to record for the purposes of producing an efficient, well-documented GIS database and for ensuring that the resources generated can be effectively archived and re-used via the ADS. For ease of use, this information is presented as a number of clearly bulleted lists within the main body of the sections but it is also crucially important to realise the cumulative nature of metadata. For example, if a practitioner's immediate concern is with the integration of a number of paper maps into a GIS, using a digitiser to create a number of vector layers, they should take careful note of the following:

- Maps and Plans (Sub-section 3.6.1)
- The Vector data model (Sub-section 3.2.1)
- Digitising (Sub-section 3.6.1.3)
- Structuring, organising and maintaining information (Section 4).

To assist this process of identifying relevant information and to ensure that adequate metadata is recorded, Section 1.3 contains a number of optimum pathways, like that illustrated above, which guide users through some of the more common GIS-related tasks and operations.

1.3 PATHWAYS THROUGH THE GIS *GUIDE TO GOOD PRACTICE*

The following sub-section contains some suggested optimum documentation pathways. These relate to tasks you will frequently undertake during the course of GIS database design, use, maintenance and archiving. In each case a routine GIS task is identified, and the relevant documentation check-lists (i.e. the bulletted lists that feature throughout the guide) are highlighted. The pathways presented here are not intended to be exhaustive nor to be viewed as a prescriptive list. Tasks frequently overlap and it is often the case within archaeology that the individuals involved in data entry are also intimately involved in the tasks of overall database management and archiving.

Practitioners are thus encouraged to use the pathways presented here as templates to develop their own 'good practice' check-lists. These can be fixed to the wall alongside computers, attached to devices such as digitisers and scanners or distributed as 'mini-guidelines' to other practitioners within their organisation.

1.3.1 Digitising a mapsheet

Here the following documentation check-lists should be consulted each time a mapsheet is incorporated into the overall GIS database:

- Maps and Plans (3.6.1)
- The Vector data model (3.2.1)
- Digitising (3.6.1.3)

1.3.2 Scanning a mapsheet

Here the following documentation check-lists should be consulted each time a mapsheet is incorporated into the overall GIS database:

- Maps and Plans (3.6.1)
- The Raster data model (3.2.3)
- Scanning (3.6.1.2)

1.3.3 Integrating an Aerial Photograph

Here the following documentation check-lists should be consulted each time an aerial photograph is incorporated into the overall GIS database:

If the photograph is to be scanned:
- Aerial photography (3.6.4)
- The Raster data model (3.2.3)
- Scanning (3.6.1.2)

If it is to be digitised:
- Aerial photography (3.6.4)
- The Vector data model (3.2.1)
- Digitising (3.6.1.3)

If heads-up digitising is to be used:
- Aerial photography (3.6.4)
- The Raster data model (3.2.3)
- Scanning (3.6.1.2)
- The Vector data model (3.2.1)
- Digitising (3.6.1.3)

1.3.4 Integrating SMR-based data

Here the following documentation check-lists should be consulted each time SMR data is to be integrated within the overall GIS database:

For the spatial component:
- Maps and Plans (3.6.1)
- Integrating textual and numeric data (3.6.2.1)
- The Vector (3.2.1) or Raster (3.2.3) data model as required

For the attribute component:
- Issues to consider when structuring and organising a flexible attribute database (4.1)
- Data standards (4.3.1)

1.3.5 Designing a GIS database

Here the following documentation check-lists should be consulted during the planning stage of the process:

- Combining and Integrating attribute databases (4.3)
- Layers and Themes (4.1)
- Choice of vector, raster or combined forms of spatial database (4.2)
- Copyright issues (4.5)

1.3.6 Routine maintenance on a GIS database

Here the following documentation check-lists should be regularly consulted:

- Sources of data (5.3.1)
- Processes applied (5.3.2)
- Dublin Core metadata (5.4)

1.4 THE THEMATIC SECTIONS

To enable practitioners to target the individual sections most relevant to the particular task at hand, the aims of each section are summarised here:

- Section 2 provides a brief introduction to the role of GIS within archaeology, containing many pointers to important core references and fundamental texts. The aim of this section is to provide a contextual background to GIS in archaeology.
- The aim of Section 3 is to discuss the principle types of primary spatial and attribute data, common sources of these data, and the processes by which they can be integrated within a GIS environment.
- In Section 4 the procedures and considerations involved in the effective structuring, organisation and maintenance of an active GIS database are discussed. In addition, suggestions for working with derived data are introduced.
- The aim of Section 5 is to discuss the importance of careful documentation. In effect *what* to record and *when*, in order to facilitate the convenient discovery and re-use of both active and archival GIS-based data resources. The concepts of documentation and metadata will be explored and discussed together with a working introduction to the Dublin Core, the metadata standard for resource discovery adopted by the ADS.
- Section 6 contains a practical discussion on how to deposit and catalogue with the ADS. This will include a detailed introduction to the creation of metadata records.
- In Section 7.1 a carefully selected set of bibliographic references is provided to enable individuals to study topics in more detail. In addition an internal glossary is presented in Section 7.2.

Section 2: A brief introduction to GIS and Archaeology

2.1 INTRODUCTION

The explosion of interest in GIS over the last ten years reflects the importance of space, spatial concepts and spatial modelling, both in the present and in the past. Although the technology of managing and analysing spatial data is now a multi-million dollar industry, which, it should be noted, pays scant regard to the specific requirements of archaeologists, our discipline has been innovative and proactive in developing its own applications of GIS. This section will serve to trace the main developments in this process, illustrating the evolving shape of GIS applications within archaeology. As will be seen, these cover a wide range of approaches and serve to emphasise the breadth and diversity of such applications within the discipline.

2.2 SOME CORE REFERENCES

Throughout the discussion reference will be made to a carefully selected set of mainstream references that should be easily accessible. The following four volumes of published conference papers (in order of conference not publication: Allen *et al.* 1990; Aldenderfer and Maschner 1996; Maschner 1996; Lock and Stancic 1995) act as a useful core framework. These provide a considerable range of case-studies and theoretical discussions together with valuable overviews of the development of GIS in archaeology (for example, Harris and Lock 1990; Kvamme 1995; Harris and Lock 1995; Maschner 1996). Another important source of references are the proceedings of the annual Computer Applications in Archaeology conference (CAA), which saw its first GIS paper in 1986. From 1992 CAA became more international and all aspects of the theory and application of GIS now form a major component of its programme (the proceedings in order from 1986 until 1995 are: Laflin 1986; Ruggles and Rahtz 1988; Rahtz 1988; Rahtz and Richards 1989; Lockyear and Rahtz 1991; Lock and Moffett 1992; Andresen *et al.* 1993; Wilcock and Lockyear 1995; Huggett and Ryan 1995; Kamermans and Fennema 1996). Two particularly useful web-based resources are GIS in Archaeology Bibliography (1995) and a list of Archaeologists using GIS which can both be found from:
http://www.archaeology.usyd.edu.au/resources/index.html.

2.3 THE EARLY YEARS AND SPATIAL STATISTICS

The first archaeological use of GIS was in North America, where it developed within the requirements of cultural resource management based on the predictive modelling of site location

(Kohler and Parker 1986). The statistics involved were well suited to raster data models and effective methodologies and results were rapidly accumulated (Kvamme and Kohler 1988; Kvamme 1990; Warren 1990). More recently there has been interest in these approaches in The Netherlands (Brandt *et al.* 1992; van Leusen 1996), and Wheatley (1996) has incorporated cultural data to overcome a major criticism concerning the emphasis on environmental data and the resultant accusations that such studies fostered an uncritical environmental determinism.

Although it has been recognised for a long time that the GIS environment is an ideal medium for the development of new approaches to spatial analysis there are very few formal statistical methods generally available (Openshaw 1991; Fotheringham and Rogerson 1994), since most commercial GIS packages lack the most basic statistical facilities. Within archaeology there is an emphasis on cell-based manipulation as an extension of the earlier work, for example auto-correlation (Kvamme 1993), statistics and simulation (Kvamme 1996), perhaps within the wider procedures of cartographic modelling (Tomlin 1990, generally; van Leusen 1993).

2.4 LANDSCAPES, PRESENT AND PAST

The archaeological awakening to GIS and the resulting rapid increase in applications started with the publication of *Interpreting Space* (Allen *et al.*) in 1990. Since then, in very general terms, there have been two streams of development which can be categorised as Cultural Resource Management (CRM) and landscape analysis. While any definition of GIS will undoubtedly emphasise analytical capabilities (Martin 1996 for an introduction), it must be recognised that a major strength of the software lies in its ability to integrate and manage large and diverse data-sets. The integration and georeferencing of different types of spatial data over large geographical areas, typically a region or even a whole country, together with textual databases is a central concern of CRM. This is usually based on statutory obligations and frequently involves the integration of data sources at a range of varying scales. As a result of the flexibility and strength of these data management capabilities it is not surprising that in the majority of cases, though by no means exclusively, analysis is relegated to a secondary role.

The potential of GIS in CRM was recognised by many countries at a conference in 1991 (Larsen 1992), and has since been realised by some of them, for example France (Guillot and Leroy 1995), The Netherlands (Roorda and Wiemer 1992) and Scotland (Murray 1995). The adoption of GIS by national and regional CRM organisations is a complex business, often embroiled within a range of concerns including information strategies (not least upgrading from an existing system), communications and standards, politics and funding. There has been a great deal of published discussion about these wider issues including European Union initiatives (van Leusen 1995), various data models (Arroyo-Bishop and Lantada Zarzosa 1995; Lang and Stead 1992) and the issues involved in the restructuring of an existing database (Robinson 1993).

While CRM systems are usually based upon a vector data model, they often need to incorporate a number of raster data layers into the database, for example the integration of aerial photographs into the Scottish National Monuments Record (Murray and Dixon 1995). Other common raster data-layers that could be encountered include the results of geophysical survey (a good example of this, although not strictly CRM, is the Wroxeter Hinterland Project

(Gaffney, van Leusen and White 1996)) and satellite imagery (Cox 1992; Gaffney, Ostir, Podobnikar and Stancic 1996).

Within the field of non-CRM landscape applications there are a considerable number that utilise the mapping capabilities of GIS rather than any of its analytical functionality. Even so, analysis can be central to GIS-based landscape studies, as demonstrated by the early case-study of the island of Hvar (Gaffney and Stancic 1991; 1992) and the seminal paper by van Leusen (1993), where the archaeological analyses engage a battery of techniques including various statistics and distance functions. Such applications served to generate a vigorous debate on the underlying epistemology of GIS and the symbiotic relationship between GIS and archaeological theory. This is a debate that raged in geography several years ago (Taylor and Johnston 1995, for an overview) and surfaced in archaeology as an argument against a return to positivism and environmental determinism (Wheatley 1993), both parts of an outdated theoretical stance long since rejected by the majority of archaeologists (Gaffney and van Leusen 1995).

Reactions to this debate have focused on attempts to integrate current theoretical notions of landscape within GIS functionality involving various ways of effectively humanising the landscape. Initially these approaches attempted to comment on the perception and cognition of an individual situated in the landscape based on visibility and intervisibility studies involving line-of-sight and viewshed routines (for example, Gaffney *et al.* 1995; Lock and Harris 1996). This resulted in the development of a new technique specifically of interest to archaeology, cumulative viewshed analysis (Wheatley 1995)

2.5 CURRENT CONCERNS

2.5.1 Theory

A more recent consideration is that meaning is culturally embedded within a landscape (Tilley 1994) and simply identifying intervisibility between monuments and places does not constitute explanation. Meaning is a multi-faceted, qualitative, measure that cannot be reached with purely quantitative tools such as GIS. This argues for the application of the technology to be theory-driven rather than data-driven as is often the case, and as part of this ongoing debate there have been two recent and quite different approaches. Llobera (1996) has attempted to formalise various indices of landscape topography and perception by writing new routines within a raster environment, in effect introducing formal methods which are embedded within a social theory of being in the landscape and of the humanisation of space. The other work, while rooted in much of the same theory (Gillings and Goodrick 1996), proposes a more phenomenological approach integrating Virtual Reality modelling with GIS, thus emphasising the importance of engagement with a locale through experiential analysis.

2.5.2 Technology

Several other themes worth mentioning are concerned with the technology of GIS, its application and functionality rather than application-specific case studies. Temporality and 3-dimensional GIS are areas that have seen relatively little work in archaeology although the early paper on archaeology, time and GIS by Castleford (1992) is still important and Harris and Lock (1996) demonstrate the potential of fully functional 3D GIS using a voxel data structure for spatio-temporal modelling of excavation data. Other topics of interest are alternative data structures (Ruggles 1992), the importance of perception surfaces, effort surfaces and time surfaces (Stead 1995), modelling ecological change (Verhagen 1996; Gillings 1995) and the potential of neural networks (Claxton 1995).

2.5.3 Intra-site studies

While there is considerable use of CAD for excavation recording and processing there is very little application of GIS, a topic which will be discussed in detail in the context of another ADS *Guide to Good Practice*. Powlesland has been a champion of integrated on-site digital recording and analysis for many years and has developed his own software (Lyall and Powlesland 1996), as has Arroyo-Bishop (Arroyo-Bishop and Lantada Zarzosa 1995). Conversely, though, Biswell *et al.* (1995) discuss the severe limitations of modern commercial archaeology in terms of integrating GIS into existing working practices while at the same time demonstrating its potential with a series of intra-site spatial analyses that highlight the difference between CAD and GIS.

Section 3: Spatial data types

3.1 SPATIAL DATA

Spatial data can be most simply defined as information that describes the distribution of things upon the surface of the earth. In effect any information concerning the location, shape of, and relationships among, geographic features (Walker 1993; DeMers 1997). In archaeology we routinely deal with an enormous amount of spatial data, varying in scale from the relative locations of archaeological sites upon a continental landmass down to the positions of individual artefacts within an excavated context. The first half of this section highlights the most important issues that need to be considered in incorporating common sources of spatial data within a GIS database. It comprises a short review of the particular issues that relate to obtaining and integrating spatial data within the GIS database. This concentrates on generic concerns such as projections, precision, accuracy and scale and is followed by a consideration of more source-specific issues. Throughout, the emphasis is upon the importance of carefully recording information about the various data themes.

3.2 PRINCIPAL GIS SPATIAL DATA MODELS

The emphasis of this guide is upon GIS as it is currently widely used within the discipline of archaeology. There are two principal GIS data-models in widespread use, which are termed *vector* and *raster*. They differ in how they conceptualise, store and represent the spatial locations of objects.

It should be noted that, to date, the principal applications of GIS within archaeology have been restricted to 2-dimensional models, and at best 2.5 dimensional representations. The latter are a result of the inability of currently available analysis and display tools to adequately deal with truly 3-dimensional data. As a direct result, the issues we will be discussing here are concerned solely with the integration, management, analysis and archiving of representations of 2/2.5-dimensional space. Issues concerning 3-dimensional spatial representations will be discussed in detail in a forthcoming CAD *Guide to Good Practice* and no doubt in summary in future editions of the present guide.

3.2.1 The Vector model

In the vector model, the spatial locations of features are defined on the basis of co-ordinate pairs. These can be discrete, taking the form of points (POINT or NODE data); linked together to form discrete sections of line (ARC or LINE data); linked together to form closed

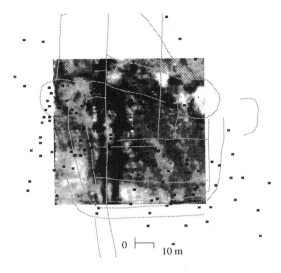

Cottam Project geophysics data displayed as a raster background and overlaid with an aerial photograph interpretation (line data), coins and metal artefacts (black point data).

Figure 1: Vector and Raster Data – Figure created by Peter Halls using data from the Cottam Project directed by Julian Richards. Image copyright © Archaeology Data Service.

boundaries encompassing an area (AREA or POLYGON data). Attribute data pertaining to the individual spatial features is maintained in an external database.

In dealing with vector data an important concept is that of topology. Topology, derived from geometrical mathematics, is concerned with order, contiguity and relative position rather than with actual linear dimensions. A good illustration of a topological map is that of the London Underground metro system. This well-known map is a precise representation of the stations (points or nodes) and the routes (arcs or lines) between them, yet provides only a very approximate indication of their relative locations and no indication of distances between them.

Topology is useful in GIS because many spatial modelling operations do not require co-ordinate locations, only topological information – for example to find an optimal path between two points requires a list of the arcs or lines that connect to each other and the cost to traverse them in each direction. It is also possible to perform the same spatial modelling and interrogation processes without using stored topology, by processing the geometrical data directly, as in such GIS as ArcView and MapInfo, or by generating topology *on the fly*, as and when it is required. The latter is the approach taken by Intergraph, amongst other major GIS suppliers.

For a detailed discussion of the vector model see Aronoff (1989) and Burrough (1986).

3.2.2 Important information to record about vector files

The following information should always be recorded when assembling, compiling and utilising vector data:

- The data type: Point, Line or Area
- Type of topology which the file contains, such as line, network, closed area or arc-node
- Details of any automatic vector processing applied to the theme (such as snap-to-nearest-node)
- State of the topology in the file, particularly whether it is 'clean' (topologically consistent) or contains inconsistencies that may require further intervention or processing. This is particularly important where arc-node data is concerned
- Projection system
- Co-ordinate system

3.2.3 The Raster model

Here the spatial representation of an object and its related non-spatial attribute are merged into a unified data file. In practice the area under study is covered by a fine mesh, or matrix, of grid cells and the particular ground surface attribute value of interest occurring at the centre of each cell point is recorded as the value for that cell. It should be noted that whilst some raster models support the assignment of values to multiple attributes per discrete cell, others adhere strictly to a single attribute per cell structure.

Within this model spatial data is not continuous but is divided into discrete units. In terms of recording where individual cells are located in space, each is referenced according to its row and column position within the overall grid. To fix the relative spatial position of the overall grid, i.e. to *geo-reference* it, the four corners are assigned planar co-ordinates. An important concept concerns the size of the component grid cells and is referred to as grid-resolution. The finer the resolution the more detailed and potentially closer to ground truth a raster representation becomes.

Unlike the vector model there are no implicit topological relationships in the data, we are after all not recording individual spatial features but instead the behaviour of attributes in space. For a detailed discussion of the raster model see Aronoff (1989) and Burrough (1986).

3.2.4 Important information to record about raster files

The following information should always be recorded when assembling, compiling and utilising raster data:

- grid size (number of rows and columns)
- grid resolution
- georeferencing information, e.g. corner co-ordinates, source projection.

3.3 GENERIC ISSUES

3.3.1 Projections and co-ordinate systems

Projection is the process by which the irregular three-dimensional form of the earth's surface is represented systematically on a two-dimensional plane, most commonly in the form of a map. Closely linked to the topic of projection is that of co-ordinate systems, which enable us to locate objects correctly on the resulting flat maps. Although we can locate objects on the globe using geographical co-ordinates expressed in units of latitude and longitude, most commonly we utilise a Cartesian or planar co-ordinate system with a fixed origin, a uniform distance unit of measure, and a pair of perpendicular axes usually termed Easting and Northing. Identification of the projection that was used in the creation of a data source is an essential first step in incorporating it into a spatial database.

For very small study areas, it is sometimes acceptable to ignore projection, and to assume that the region of interest is comprised of a flat, two-dimensional surface. However, if the study region is larger than a few kilometres, or if information is to be included from data sources, e.g. mapsheets, which have been constructed with different projections, then a GIS needs to understand the projection used for each layer in order to avoid inaccuracies.

Projection consists of two main stages: first the surface of the earth is estimated through the use of a geometric description called an ellipsoid (sometimes, though not always correctly, referred to as a spheroid), and secondly the surface of this ellipsoid is projected on to a flat surface to generate the map. Ellipsoids are defined in terms of their equatorial radius (the semi-major axis of the ellipse) and by another parameter, such as the flattening, reciprocal flattening or eccentricity. However, as a user of GIS, the ellipsoid definition is usually uncomplicated to incorporate. Most of the ellipsoids which have been used to generate maps have names such as the 1830 'Airy' spheroid (used by the Ordnance Survey) or the 'International' or 'Hayford' ellipsoid of 1909, and it should be sufficient to provide the full name of the ellipsoid used (for a simple introduction to ellipsoids see Defence Mapping Agency 1984).

3.3.2 Projection methods

There are a huge variety of methods available for undertaking the projection itself. Since by their very nature projections are a compromise, each method produces a map with different properties. In a cylindrical projection, for example, the lines of latitude (parallels) of the selected ellipsoid are simply drawn as straight, parallel lines. In the resulting map the parallels become shorter with distance from the equator, and to maintain the right-angled intersections of the lines of latitude and longitude, the lines of longitude (meridians) are also drawn as parallel lines. This maintains the correct length of the meridians, at the expense of areas close to the poles which become greatly exaggerated in an east-west direction. A transverse cylindrical projection is created in the same manner, but the cylinder is rotated with respect to the parallels and is then defined by the meridian at which the cylinder touches the spheroid rather than the parallel.

The Mercator projection exaggerates the distance between meridians by the same degree as the lengths of the parallels, in order to obtain an orthomorphic projection. A transverse Mercator is similar, but based on the transverse cylindrical projection. There are also many other forms of projection which are not based on the cylinder, including conical projections (based on the model of a cone, placed with its vertex immediately above one of the poles) and entirely separate families of projections such as two-world equal area projections and zenithal projections.

Depending on the projection used, different parameters will need to be specified in order to define it. Basic projections are often modified through the use of correction factors. In transverse projections, for example, it is not uncommon for a scaling factor to be applied to the central meridian to correct for the east-west distortion of the projection itself. A projection may also use a false origin, which arbitrarily defines a point on the projection plane to be the point 0,0. False origins are normally used to ensure that all co-ordinates in the area of the projection have positive values.

All projections have limits beyond which one or more of their attributes will become too distorted. For example, the Transverse and Universal Transverse Mercator projections work well only for a narrow east-west width – around 6 degrees of longitude – beyond this limit the distortion increases rapidly. When choosing a map projection it is essential to check the details of both the capabilities **and** the limitations of any given projection method against the nature of the area of interest: size, extents, nature of use, etc.

Details of projection procedures can be found in a variety of standard texts, for example Bugayevsky and Snyder (1995), Snyder (1987) and Evenden (1983; 1990). Standard software for specifying and undertaking cartographic projection is available for a variety of platforms. Probably the most flexible is the PROJ 4 product of the USGS discussed by Evenden (1990).

3.3.3 Co-ordinate systems

Geographical co-ordinates, expressed in angles of latitude and longitude are used to locate features upon the globe whereas planar Cartesian co-ordinate systems are used to locate features upon projected maps.

In a planar co-ordinate system, the relative positions of objects represented upon a given mapsheet can be specified using standard units of distance measured with respect to a fixed origin point. The precise characteristics of a given co-ordinate system depend heavily upon the projection used to generate the 2-dimensional representation; as a result coordinate systems are as numerous as projection systems.

For example, there are five primary coordinate systems in use in the US, a country with a very broad east-west extent. Some of these are based upon the properties of specific map projections and others on historical land division strategies (DeMers 1997: 63).

By contrast, a single planar co-ordinate system based upon the transverse Mercator projection is used for all recent mapping for Great Britain, which is elongated north-south but narrow east-west. The British National Grid uses the Airy spheroid for its datum, and its origin is located at 49 degrees North, 2 degrees West. There is a false origin defined at 400,000, -100,000 such that the central meridian is the Zero *easting* of the National Grid – which is also aligned to the 2 degree West meridian of Longitude. Additionally, the scale along the central meridian is 0.9996012717 (as opposed to the more normal scale of 0.9996 for the Transverse Mercator projection): this results in distances along the northing 180,000, almost exactly half way across Great Britain, being to true scale. The units of the National Grid are metres measured east and north from the origin. A similar Transverse Mercator projection, but with an origin further to the west, is used for Ireland.

It is important to realise that projections and the resultant planar co-ordinate systems vary across nations and through time. The British National Grid has only been in use since the 1940s. British mapping prior to this time was based upon the Cassini projection and used an independent datum for each *county*. This mapping is often termed the *County Series*. A problem common to all situations where multiple datums are in use for any one country is that features which cross the borders of the different datums may not match when the respective map sheets are brought together. This problem certainly exists in the pre-war British mapping and is also present in the State-Plane system in use in the US today.

As with the British Ordnance Survey, many national mapping agencies use local projections designed to suit the size and shape of the area covered by their maps.

The Universal Transverse Mercator (UTM) system is a planar projection where degrees of longitude and latitude form a rectangular grid. Since distortion tends to increase most markedly on either side of the central meridian with this projection, UTM is used for narrow north-south oriented zones.

The world is divided into zones each covering 6 degrees of longitude and numbered in an easterly direction from 1, centred on 177 degrees west, to 60, centred on 177 degrees east. Within each zone, a transverse Mercator projection is established with its origin at the intersection of the central meridian with the equator. The false origin is offset so that the central meridian is at 500,000 metres east. The false Northing is zero metres in the northern hemisphere and 10,000,000 metres in the southern hemisphere. The scaling factor on the central meridian is 0.99960.

Despite its world-wide applicability, UTM has some disadvantages. Different ellipsoids may be required in different parts of the world, and transformations between zones are required when the area of interest covers more than one zone. The UK, for example, is mostly in zone 30, but areas east of the Greenwich meridian fall into zone 31. The US State-Plane co-ordinate system is an application of UTM.

The use of three-dimensional absolute co-ordinates from satellite positioning systems introduces further complications. These systems measure the relative positions of receiver and satellites using an 'Earth-Centred, Earth-Fixed' Cartesian co-ordinate system (ECEF). This system, which is aligned with the World Geodetic System 1984 (WGS 84) reference ellipsoid, has its origin close to the earth's centre of mass, its z axis parallel with the direction of the conventional terrestrial pole and its x axis passing through the intersection of the equator and the Greenwich meridian. Fortunately most receivers convert ECEF co-ordinates to WGS 84 latitude, longitude and height for output, and some will also perform transformations to other datums and co-ordinate systems, for example to the Great Britain National Grid.

3.4 PRECISION AND ACCURACY

In incorporating any spatial data source it is crucially important to consider the issues of *precision* and *accuracy*.

> "Precision implies that the degree of measurement of an attribute is refined; accuracy that the measurement taken is correct within the degree of precision indicated" (Richards and Ryan 1985: 20).

This can best be illustrated with respect to the example of a highly detailed topographical survey undertaken using a total-station survey instrument and based upon reference points taken from a 1:2500 scale base-map. This is an example of a potentially very ***precise*** method using potentially very ***inaccurate*** data. This is because the source data, i.e. the fixed point locations derived from the map, are not sufficiently accurate to justify the precision employed in the method. To summarise, accuracy relates to the correctness of a result, whereas precision is essentially a measure of the units used.

As an aside, such issues of precision and accuracy are of particular interest in the context of GPS, where the quoted accuracy of co-ordinates varies commonly from sub-centimetre to around ±50 or so metres. Although this varies according to where on the globe you are referring, one second is approximately equal to thirty metres, yet, in terms of decimal degrees, is represented by a value of 0.0002777778. Thus a typical GPS readout in decimal degrees of, say, 52.005N is only within about 500 metres – making the general GPS error rate of ±50 metres trivial!

There is a temptation to believe the apparent accuracy with which computer-based GIS report the co-ordinate locations of objects. This information is, however, only as good as the initial data source, and errors made at this stage can be perpetuated through the lifetime of the data set.

3.5 SCALE AND RESOLUTION

Scale is the ratio of the distance measured on a map to that measured on the ground between the same two points. For example a quoted scale of 1:50,000 implies that a distance of 1 cm on the map translates to a distance of 50,000 cm (or 500 metres) on the ground. Often, the difference between large and small map scales is confused. The larger the ratio, the smaller the map scale. Therefore, a map of the world would have a very small scale, whereas a map of a town centre will have a large scale.

Resolution is the smallest distance that can be usefully distinguished on a map with a given scale, for example on a 1:10,000 scale map the smallest distinguishable distance is 0.5 mm which equates to a distance of 5 m on the ground. It is worth noting that the accuracy of a map cannot be 'better' than its resolution, but it can often be much 'worse'.

The larger the map scale, the higher the possible resolution. It is very important to be aware of the scale of a given spatial data source as the degree of simplification and reduction involved in the representation of spatial features tends to increase as scale decreases. As map scale decreases, resolution diminishes and feature boundaries must be smoothed, simplified, or not shown at all. This process is referred to as generalisation. To give an arbitrary example, a map of an area of rural Greece produced at a scale of 1:5000 may show villages and towns as discrete areas, whereas at a scale of 1:500,000 they will be portrayed as little more than dots.

It should also be noted that the wider usability of any co-ordinate system is partly a function of the resolution in which it is quoted. For example, the widely used six figure OSGB grid references are only valid within the hundred kilometre grid square for which they are quoted – they repeat in every other hundred kilometre grid square and they also only address to the nearest kilometre, even though the OSGB National Grid per se can support references to the nearest metre (or even sub-metre, if given as a decimal number).

In a GIS where spatial data sets from a range of sources are integrated and the spatial resolution of a given data set can be altered at will, it is vitally important to be aware of such issues and not to analyse spatial information at a scale greater than that of the data source (see DeMers 1997: 56 for discussion).

3.6 COMMON SOURCES OF SPATIAL DATA

In undertaking any GIS-based work the most common sources of spatial data will comprise one or more of the following:

*   mapsheets and plans
*   raw co-ordinate lists, derived from field survey or extracted from existing site records such as those held within national and regional monument records
*   aerial photographs and remotely sensed images
*   digital data products, such as the United States Geological Survey topographical data.

3.6.1 Maps and plans

Mapsheets comprise one of the most widely available and familiar sources of spatial data. In incorporating spatial data derived from mapsheets it is important to be aware of a number of issues. The first of these concern the map itself. The medium of the map itself is highly important. While maps originated on specially stable plastic films, such as mylar, are reasonably stable, paper maps can stretch and distort over time. In addition, where the map is a copy rather than an original a number of distortions may be present as a result of the specific copying process used. In general the following information should always be recorded:

*   Publisher and copyright owner, which will often (but not always) be the same. For Ordnance Survey mapping, the copyright holder is the Crown.
*   The map medium.
*   Scale of source map, given as a ratio, and the original scale (where the source map is an enlargement or generalisation from another map).
*   Name of the map and the map series (where appropriate).
*   Claimed accuracy for any specific map components: map makers will often provide an estimated precision for contour lines or other sub-components of a map.
*   All details of the map projection and co-ordinate system employed. This information is usually printed on the mapsheet or else should be sought from the map source.

3.6.1.1 Integrating map data

There are three methods for integrating map data into a GIS database and these are based upon two discrete techniques. To re-affirm the point raised in 1.2, where information to be recorded is indicated, this should be considered as being in addition to the generic information required for all mapsheets, as described above.

3.6.1.2 Scanning

Paper mapping can be scanned, with a flatbed or drum scanner, to generate raster GIS data themes. Scanning devices vary considerably in accuracy and resolution, with flatbed and drum scanners normally providing a resolution between 100 and 1200 dots per inch (dpi). The more expensive drum scanners claim resolutions of between 3–5000 dpi. In all cases care should be taken to distinguish between the true optical resolution of a given scanner and that obtained through interpolation procedures. If scanned, then there is likely to be a single raster file data product.

There are a very wide variety of image formats for holding raster data (e.g. TIFF, GIF, JPEG), the majority of which are designed for photographic images and not spatially referenced data. Several GIS provide proprietary raster data structures and record spatial referencing information (e.g. IDRISI, Arc/Info GRID, SPANS raster, GRASS raster), they also provide tools for importing data from other common raster formats. More recently the Tagged Interchange File Format (TIFF) graphics standard has been extended to provide georeferencing and spatial data in a format called 'geotiff'. Details of the geotiff standard, including the official specification of Geotiff 1.0 can be obtained from the Geotiff WWW page. Although currently supported by a limited number of proprietary GIS, many manufacturers have committed to supporting the standard which should provide a platform-independent method for archiving and transferring spatially referenced raster products.

It should be noted that the scanning process can result in some very large raster images and this can be compounded by the software used to integrate and study the raster layers which may require increased colour depth.

For products that have been generated by scanning paper originals, the following additional information to the core mapsheet data should be recorded for each raster file generated. It should be noted that to retrieve some of this information will involve careful checking of the respective hardware and software documentation:

- Details of the scanning device used, such as the make and model, software driver and version
- Parameters chosen in the scanning process, such as the resolution setting of the device, the number of bits per pixel used
- Details of any pre-processing undertaken on the source mapsheet. This may include a range of options provided by the specific scanning software used
- Details of any post-processing undertaken on the data, such as noise reduction or sharpening with convolution filters, histogram equalisation, contrast adjustment

3.6.1.3 Digitising

Maps and plans may also be geometrically described, using a digitising tablet, to provide vector data. Digitising tablets generally offer finite resolution in both x and y directions. This can be expressed as a quoted resolution, for example 0.02 inches or 0.001 inches, or as lines per inch (lpi), e.g. 200 lpi or 1000 lpi. This information can be found within the digitiser manual. Unlike the scanning process, where a scanned map generates a single raster GIS image, digitising a single paper map may form the basis of a large number of discrete, thematic vector data layers.

When digitising mapsheets the following additional information should be recorded. As with the scanning process this may involve careful checking of hardware and software documentation, for example to determine the resolution of the digitiser.

- Detail of the digitising device used, such as the make and model, software driver and version

- The precision, usually specified as a quoted resolution or as lpi
- Details of any automatic vector processing applied to the theme (such as snap-to-nearest-node)
- Details of control points used to manage conversion from digitiser to real-world planar co-ordinate systems
- Errors incurred in the above transformation process (e.g. quoted RMS)

3.6.1.4 Scanning-digitising hybrid

A third option is to scan the source document but then use the scanned product as the basis for 'on screen digitising', using a graphics workstation and pointing device to create vector data themes. This is often referred to as 'heads-up digitising' and is an attractive option where a digitising tablet is not available, or where raster data from a third party can be obtained.

The basis of heads-up digitising is to use the mouse pointer to trace around the image to be digitised, recording the coordinates as it moves – in much the same manner as moving the puck on a digitising table or tablet. With the image displayed on the computer screen, however, it is easy to *zoom in* on an area of complexity in a way that is hardly possible on the digitising table or tablet. Indeed, with the scanned image, the only limit to this enlargement is the point at which the individual cells, representing the marks on the original, are distinguishable as squares and rectangles. Where the scan is at 300 pixels per inch, typical of many desktop scanners, each of these cells represents approximately 0.085mm square; with higher resolution scanners the size of these cells is proportionately smaller! In addition to this facility, because registration is effectively performed when the map/plan is scanned, it is much easier to undertake the digitisation in much smaller time-chunks, thus minimising errors resulting from fatigue, etc.

There are a number of software tools available to assist in obtaining vector data from a scanned image of a map or plan. These include very sophisticated, semi-automatic, tracing tools which, for an *ideal* image, can often vectorise perhaps 70–80% of the data without intervention and which automatically request intervention when a problem cannot be resolved. Examples of this type of tool include *Vtrack*, from Laserscan, and *ArcScan*, from ESRI. Tools such as these can manipulate output from high (over 3000 pixels per inch) resolution drum scanners, as well as the output from a desktop flatbed scanner. Such software tends to be expensive, although sometimes available to non-profit research and educational institutions at discounted rates. There are also, at the other end of the price/sophistication range a number of cheap/*shareware* tools for running on a PC. These may have limitations in terms of the maximum scan resolution they can handle, or the maximum size or complexity of the image. Note that **none** of these tools can be guaranteed to be able to vectorise 100% of a scanned map/plan without intervention. The degree of intervention required will always be a function of the sophistication of the vectorising/tracing tool, the quality of the scan, and the nature of the original.

3.6.2 Textual and numeric data

Often spatial data will be encoded in the form of co-ordinate lists, for example those commonly found within regional and national monument registers. Where co-ordinates are expressed they should conform to the standard Surveying notation of Easting, Northing, elevation (x, y, z) though this may not be consistently applied in proprietary systems and particularly in hand-written records.

It is important to determine how the co-ordinates were derived, for example are they reckoned from a base-map or determined through field survey? In addition, it is also important to determine the precision of the co-ordinates as quoted. For example, within regional monument registers in the UK it is not uncommon to find the locations of archaeological sites quoted to the nearest 100 metres (what is referred to as a 6 figure grid reference). If this data is to be integrated into a GIS database comprising spatial data originated at one metre resolution, these co-ordinates will have to be rounded up, leading to a spurious level of accuracy.

One important point to realise when using co-ordinate references is that even when a discrete point reference is quoted it is in actuality indicating the lower left-hand corner of a bounding box. The size of this bounding box is dictated by the resolution of the reference. To return to the example of a UK sites and monuments register, far from indicating the precise location of a site on the ground, the six figure reference actually serves to locate the bottom left-hand corner of a 100 x 100 metre bounding box, somewhere within which the site is located.

3.6.2.1 Integrating textual and numeric data

When integrating textual and numeric data it is important to understand the co-ordinate system to which the quoted co-ordinate locations relate. Although co-ordinates will most commonly reference a national or international system, such as the Great Britain National Grid, or UTM, occasionally they may relate to a site-grid. This is a contingent rectangular co-ordinate system with its datum at a fixed point, that has been established for a specific purpose. Perhaps the most common examples are geophysical survey grids and excavation grids, established to facilitate the spatial recording of features within excavated contexts. Until the location of grids has been surveyed or 'fixed' with respect to a larger or more generic co-ordinate system, such as those mentioned above, it can be thought of as 'divorced' or 'floating' and whilst internally consistent is impossible to relate it spatially to features beyond the confines of the grid. Needless to say, the geo-referencing status of grids must carefully be considered when archiving data for potential future re-use as this can seriously affect the overall accuracy of the quoted co-ordinates. For a detailed discussion of these issues undertaken in the context of geophysical survey, practitioners are referred to the ADS *Guide to Good Practice* in Archaeological Geophysics.

When integrating textual and numeric data the following information should be recorded:

- The data source
- The precision of the quoted co-ordinates
- Have the quoted locations been verified and how?
- Projection system/co-ordinate origin
- If derived from a source map, where possible record details of the map-base used (see the paragraph on map data for details of the information required)
- If derived from a survey programme, where possible record details of the survey procedure (see the paragraph on Survey data for details of the information required)

3.6.3 Purchased or downloaded digital data

Spatial data which is already in digital form may be purchased from mapping agencies (such as the Ordnance Survey or public utilities). Many agencies supply both raster and vector data, depending on the requirements of the user. An increasing amount of spatial information can also be downloaded from the Internet, again in both vector and raster formats.

A useful list of digital data sources can be found in Appendix 1.

3.6.3.1 A note on the integration of digital data sources

Looking to the information that it is important to record, it must be realised that such data is often derived from another medium, for example a scanned or digitised map or a scanned image of a geophysical survey. As a result, similar information to that required for these data sources should also be recorded for digital data products. This information will be obtainable directly from the supplier and should be requested if not supplied.

As discussed in 3.2.1, vector data may take the form of simple points or lines, often with associated attributes, or more complex topological themes such as arc-node data. Because of the variety of data structures used in different GIS, particularly for arc-node data, there is currently no platform-independent standard file format for spatial data. Several standard formats are of interest, however, and may be used in particular circumstances.

- British Standard 7567 (the National Transfer Format) is the format used by Ordnance Survey for the supply and transfer of digital products. It allows both spatially referenced raster and vector products to be stored in ASCII coded form. A useful guide to the OS implementation of BS 7567 (NTF 2.0) may be obtained from the Ordnance Survey
- Users and those involved in the archiving of spatial vector and raster data in the United Kingdom should also be aware of the National Geospatial Data Framework (NGDF), which is "a national forum of data providers and data users seeking to facilitate and encourage widespread use of geospatial data which is 'fit for purpose'. Its objectives are to facilitate and encourage collaboration in the collection, provision and use of geospatial data; to facilitate and encourage the use of standards and best practice in the collection, provision and use of geospatial data and to facilitate and widen access to geospatial data"
- SDTS (Spatial Data Transfer Standard) is a United States Federal Information Processing Standard (FIPS) which was developed to accommodate different data models to allow users to encode spatial data in a standard format, accompany data with description and provide machine and platform independence. SDTS is the responsibility of the Federal Geographic Data Committee (FGDC). SDTS is not an exchange format for data, rather it is a standard set of guidelines which will describe and preserve a database design and its underlying model. Documentation on SDTS is available from http://mcmcweb.er.usgs.gov/sdts/
- DLG (Digital Line Graph) format is used by the United States Geologic Survey for supply of vector information, while DRG (Digital Raster Graphics) is the description that the USGS gives for the distribution of scanned map sheets. Details of these standards may be obtained from the USGS WWW site.
- DXF (Drawing eXchange Format) format is commonly used for transferring drawings between Computer Aided Design systems. It is also however very widely (mis)used as a

de facto standard for the transfer of digital spatial data (Walker 1993). A detailed discussion of DXF is undertaken in the CAD *Guide to Good Practice*.

More information about standards, including spatial standards, is available in Appendix 2.

3.6.4 Aerial photography

Aerial photographs may reveal archaeological sites directly, where they are extant, or as crop, soil or other surface indications where the site is buried. As a result, archaeology has a long history of using aerial photographs for recording existing site morphology, and prospecting for new ones.

Two types of aerial photograph are widely used in archaeology: vertical photographs and oblique photographs. In either case, the image will pass through at least two stages before it can be included in a GIS database. First the photographic image will need to be scanned and then it will need to be rectified and georeferenced. For a detailed bibliography and for a full and comprehensive discussion of the issues and techniques involved, including more advanced techniques such as photogrammetry, please refer to the ADS *Guide to Good Practice* in Archiving Aerial Photography and Remote Sensing Data.

To incorporate scanned and rectified aerial photographs into GIS databases the following information should be recorded:

*   Full Photographic details
*   Details of the scanning process (see the paragraph on map scanning for details of the information that should be recorded)
*   Details of the rectification method(s) used
*   The software employed including, where possible, specific parameters chosen
*   Details regarding the ground control points (GCPs) used during the procedure
*   Details of any post-processing undertaken on the data, such as noise reduction or sharpening with convolution filters, histogram equalisation, contrast adjustment etc.

3.6.5 Satellite and airborne remote sensed images

Airborne remote sensing refers to situations in which an aircraft carries an electronic sensor that records information directly to digital format. In recent years, increasing use of remote sensing satellites has been made with images being available at reasonable cost. For a detailed bibliography and for a full and comprehensive discussion of the issues and techniques involved please refer to the ADS *Guide to Good Practice* in Archiving Aerial Photography and Remote Sensing Data.

Looking to the information that needs to be recorded, many of the issues which arise when using scanned aerial photography will also be relevant to the integration of airborne remote sensed data. For example, they will normally require rectification in the same way as scanned photographic material. Once again, comprehensive details can be found in ADS *Guide to Good Practice* in Archiving Aerial Photography and Remote Sensing Data.

*   Data source
*   Date image was captured

- Data resolution
- Details of any post-processing undertaken on the data, such as noise reduction or sharpening with convolution filters, histogram equalisation, contrast adjustment etc.
- Details of the rectification method(s) used
- The software employed including, where possible, specific parameters chosen
- Details regarding the ground control points (GCPs) used during the procedure

3.6.6 Primary survey data

3.6.6.1 Terrestrial survey

Data from older optical instruments will normally have been recorded and even processed entirely by hand. Most modern total-station and satellite-based instruments (using either US 'Navstar' Global Positioning System (GPS), or the Russian GLONASS) have internal data stores and processors, or are used with a separate data logging device, typically a hand-held or laptop computer.

Data may be obtained directly from these survey instruments, usually in the form of co-ordinate pairs (or 3D triples) often with attached attribute(s). It may be exported either to proprietary data file formats, or to ASCII files which can be imported directly to GIS databases. In many cases, there is unlikely to be a complex thematic arrangement of data, although this is changing with the increasing use of advanced data logging software that provides direct GIS or CAD input in the field.

Commonly, however, survey data will be in the form of CAD drawings, which may have thematic (layer) structure or complex block-attribute structure themselves. Here, the source and derivation of the data used to construct each layer must be documented. Readers should consult the CAD *Guide to Good Practice* to familiarise themselves with CAD systems.

Whatever the source of the data, it is essential to understand the sources of errors and to record details of any instruments, software and methods used to derive the co-ordinates used in GIS layers. Whilst modern semi-automated survey instruments and methods may be easier to use and reduce the possibility of simple transcription and mis-calculation errors, they are still subject to many sources of error. For terrestrial survey methods, these include the reliability of locations used as survey base stations as well as individual and cumulative measurement errors introduced by the instrument and its operators (Clancy 1991).

3.6.6.2 Integrating terrestrial survey data

Survey information may well be in the form of angle and distance measurements. Although there are GIS which can store and manipulate such geometric measurements, they are most often processed to derive Cartesian co-ordinates. Much in the same way as for quoted co-ordinate lists and textual data, when incorporating terrestrial survey data into a GIS database it is critically important to understand fully the co-ordinate system to which the quoted co-ordinate locations relate. Most surveys begin their lives divorced or floating, i.e. quoting co-ordinates with respect to a highly contingent rectangular co-ordinate system. These divorced surveys are internally consistent and commonly employ very precise technology, for example total-station survey instruments, capable of recording locations to the nearest millimetre and beyond. Before the results of such surveys can be more widely employed they have, however, to be integrated within a larger or more generic co-ordinate system, such as a national survey grid or UTM. Despite the high precision of such surveys, this process of geo-referencing can severely affect the overall accuracy as often the divorced survey grid is 'fixed' with respect to points derived from base-maps which themselves may only, at best, have been located to the nearest metre. The more accurate the points utilised to fix the divorced survey grid within a larger co-ordinate system, for example the use of triangulation pillars or high precision GPS, the greater the accuracy of the resultant survey data resource.

When integrating data themes which are derived from survey data, the following should be recorded:

- The source (paper/digital map, GPS, data from mapping agency) and estimated error of survey base station co-ordinates
- Details of the survey, including date time and purpose
- Details of the thematic organisation of the survey
- Make and model of instrument used
- Type of survey (contour, feature etc.)
- Estimated error terms for the co-ordinate pairs and (if appropriate) the z-co-ordinate
- Georeferencing information, overall accuracy of the survey data

3.6.6.3 Satellite-based (GPS) survey

Satellite-based survey data is complicated by the variety of possible methods used by receivers to produce a position fix, and by differential techniques used to improve the accuracy of fixes. Whilst a single fix from a simple hand-held device intended for navigational purposes may only be accurate to within 100m, differential correction may improve this to 10–15m, or better. On the other hand, under favourable circumstances, the best survey instruments may achieve sub-centimetre accuracy.

The accuracy of fixes from the same equipment varies through time as the relative positions of satellites and receiver change. Careful 'mission planning' is essential to avoid times when the satellites offer a poor configuration for triangulation or when a reduced number of satellites are visible from the observation site. Fortunately, many data logging packages include satellite prediction facilities that can be used to determine the optimum observation times. Jan van Sickle (1996) and David Wells (1986) provide useful introductions to satellite surveying, whilst Leick (1995) gives a thorough coverage of the underlying technology and mathematics.

3.6.6.4 Integrating GPS data

As discussed earlier, satellite systems measure the relative positions of receiver and satellites using an ECEF Cartesian co-ordinate system. Whilst positions expressed in ECEF x,y,z co-ordinates are ideal for locating satellites and receivers in 3D space, they are rather less suitable for terrestrial mapping. Fortunately, most receivers will output co-ordinates expressed in latitude and longitude relative to the WGS 84 ellipsoid, and many will also generate positions relative to other ellipsoids and in other co-ordinate systems such as UTM or the various national grids. If the only co-ordinates available are relative to WGS 84 or some other system, these will require transformation to the system used for the base mapping in the GIS. This conversion requires both a transformation between ellipsoids and a datum shift. Numerous datums have been used for mapping around the world. For details, see Snyder (1987; 1989). Each datum is defined in terms of a common reference ellipsoid together with x,y,z offsets that take account of differences in the origins of the ellipsoids and local deviations of the earth's surface from the ideal ellipsoid.

For mapping in the UK, the Ordnance Survey publishes two booklets (1995; 1996) giving details of transformations between ECEF, WGS 84, Latitude/Longitude, and the National Grid. Similar transformations are used to derive UTM co-ordinates.

Some GIS provide suitable datum transformation functions and several purpose-written programs are also available for this task. It is important to realise, however, that the results produced by many transformation methods are only approximations that may degrade the accuracy of the original position. For example, the methods described in Ordnance Survey (1996) produce National Grid co-ordinates that are only correct to within 2 metres. The programs and methods used in any transformation must therefore be recorded as part of the history of the data.

In integrating satellite data the following information should be recorded:

- The method used to locate stations: C/A or P code pseudorange measurements, carrier phase measurements and whether a single measurement or averaging (include time period) was used
- The software used for any co-ordinate transformation and associated error estimate
- The satellites used in obtaining fix and observed GDOP (Geometric Dilution of Precision, a measure of the quality of the fix indicating the suitability of satellite positions for triangulation)
- The nature of any differential correction undertaken together with error estimates
- The broadcast differential: name of the service provider and the name and location of base station
- The local base station: instrument details, location (including error estimate) of base station
- Post-processing: the software used and the source of correction data

3.6.6.5 Preferred and accepted formats for GPS data

GPS or GLONASS data will often have been recorded using data logging equipment and then transferred to other systems using a simple ASCII text, DXF or proprietary GIS or CAD format. For most purposes, one of these formats, particularly if it includes associated attribute data, will be preferred.

The direct output of many of these receivers is usually in one of the following formats:

- NMEA 0183: an ASCII protocol devised by the US National Maritime Electronics Association for marine navigation equipment (NMEA 1995)
- RINEX version 2: Receiver INdependent EXchange format (Gurtner and Mader 1990; Gurtner 1994)
- A proprietary ASCII or binary format such as Trimble Standard Interface Protocol (TSIP)

Of these, RINEX is widely used and is not tied to a particular device or class of device. It also has a provision for recording comments and events, such as movement to a new survey point and the start of a new point occupation. Where raw satellite data forms part of a data set, it is the currently preferred format.

3.7 ATTRIBUTE DATA – INFORMATION ABOUT THE SPATIAL FEATURES YOU HAVE RECORDED

Information commonly stored, or manipulated, using a GIS tends to have two main components – the *spatial* and the *descriptive* attributes. For many users, and with many software products, these two data types may appear to be a seamless unit. There are, however, some data management issues which are peculiar to whether you are working with spatial or attribute data, and certain general issues which are common to either form of information.

Attributes are data that describe the properties of a point, line, or polygon record in a Geographic Information System. For example, imagine a GIS coverage in which points represent sites on a landscape. The attribute data that accompanied this coverage would record more detailed information about each site. Attribute information might include an indication of the time period in which the site was occupied (e.g. Neolithic, Iron Age, Medieval), full descriptions of the archaeological deposits excavated from each site, and an indication of the class of artefacts found on the surface at each site.

Archaeological attribute data already exists in myriad forms. These can range from simple card indexes – for example the results of a graveyard survey undertaken using the Council for British Archaeology guidelines – through to complex digital databases recording a wealth of detailed information. Such databases sometime include descriptions about all the archaeological sites in a country or county and sometimes contain very detailed site-specific information such as stratigraphic records. This diversity is on the increase as the use of computers grows in archaeology.

In archaeological GIS you will often be linking and combining attribute information collected by others, and turning this information to new purposes.

3.8 COMMON SOURCES OF ATTRIBUTE DATA

Below are some likely sources of attribute data which you may come across, and wish to re-use:

- paper based card indexes
- archaeological site and survey archives (including paper based records, finds databases)
- qualitative report texts and articles published in journals (paper based or on the Internet)
- microfiche archives
- geophysical interpretation data derived from interpreted geophysics plots
- aerial photograph interpretations which may include morphological analysis, attribute data and photo source information
- typological databases or artefact type series
- data generated at a regional level for integrated large scale historic landscape studies, such as the English Heritage Open Fields Project
- local level archaeological databases (e.g. Sites and Monuments Records or Urban Archaeological Databases where they are held separately)
- local museum site and finds databases
- local Record Offices

- national archaeological databases (such as the various National Monument Records or English Heritage's database of Scheduled Ancient Monuments)
- Gardens Trust surveys
- historic buildings surveys and databases maintained by local authorities
- metadata relating to data sets

3.9 DESIGNING A NEW ATTRIBUTE DATABASE

Whether you are using pre-existing attribute data or actually collecting new information yourself, you will need to think carefully about the design of your new attribute database. A great deal of literature exists on this complicated topic – it is quite literally a topic which has launched a thousand PhD research projects! Some good sources of basic information can be found in Batini *et al.* (1992), Date (1995), Ryan and Smith (1995), and Whittington (1988).

Archaeologists should also be aware of MIDAS, the Monument Inventory Data Standard (RCHME 1998). This standard is designed for those establishing a new attribute database in which to manage archaeological information or for those who have been working with archaeological attribute databases for a long time.

3.9.1 An introduction to the principal types of database structure

Database systems should be efficient tools for the storage, analysis and reporting of your data. As a result, the choice of database package and data structure used in a given project should be dictated by the requirements of each organisation. It is not within the scope of this guide to enter into a discussion of the merits and failings of software packages. Instead a short overview of the types of database data models is presented.

Data structures currently fall into four major types: *flat file*, *hierarchical*, *relational*, and *object oriented*. More detailed discussion of these can be found in *Fundamentals of Spatial Information Systems* (Laurini and Thompson 1996), especially pages 620–38 on object oriented databases.

3.9.1.1 Flat file data structures

In this simplest form of data structure, data are arranged in concurrent horizontal rows, with attributes stored in vertical columns. One row stores all attributes for a single entry (object) on the database. If many of the objects on the database have the same attributes they must be entered many times, leading to data redundancy and often to empty fields (resulting in wasted computer resources). A common example is the card index.

3.9.1.2 Hierarchical data structures

Hierarchical data structures have useful applications within archaeology as they arrange the objects in a database in a related tree of linked parent and child records. This can be used to model the breakdown of the historic environment into 'monuments within monuments' and allows flexible searching across the hierarchy. The most common applications for these database structures are in cultural resource management environments such as SMRs and national databases which contain large amounts of data and need efficient, speedy, searches.

3.9.1.3 Relational data structures

Currently the most common type of database used is based on the relational data model. If you imagine a series of tables similar to small flat file databases, with links or relationships between specific unique fields that allow complex queries of different data sets, you have the essence of the relational structure. One table could be of pottery types, another of contexts, a third of scientific dates and with easily structured queries it will be possible to construct chronologies based on pottery typology or scientific dating.

3.9.1.4 Object oriented data structures

The newest form of data structure, there are currently a limited number of GIS packages using the object oriented approach (e.g. Smallworld). While the relational data structure deals with an object's description by tearing it apart into single rows, and holding those rows in many discrete but linked tables of similarly grouped attributes, the object oriented approach to data structure allows the descriptive attributes of an object (e.g. a monument) to be encapsulated digitally in one place, allowing a more realistic model of the 'real world' to be assembled. The geographic location of the object is then just another characteristic of the object, just as function, date and period of existence are.

3.10 ISSUES TO CONSIDER WHEN STRUCTURING AND ORGANISING A FLEXIBLE ATTRIBUTE DATABASE

When attempting to structure and organise a flexible attribute database the following factors are of critical importance. In the following section each of these issues will be looked at in turn.

- Naming conventions
- Key fields
- Character field definitions
- Grid references
- Validation
- Numeric data
- Data entry control
- Confidence values
- Consistency
- Documentation
- Dates

3.10.1 Naming conventions

Try to keep field names descriptive rather than cryptic. The crib sheet for decoding cryptic names may easily get lost, and your fields are likely to be too numerous for you to remember their contents easily.

3.10.2 Key fields

Key fields are the most important fields in your attribute database and are the fields that will be used for primary searching of the database and/or for linking tables within your database. It is essential that the same data definitions are used for all instances of the key field in your database and that the same codes are used in each.

3.10.3 Character field definitions

Take care with character field definitions. Most databases require character data to be stored in a fixed length form and so, inevitably, this means that every record must contain enough space for the largest expected, even where this is not required for the vast majority of records. As an example, there is no point in defining a location name field large enough to store the longest name in Monmouthshire, *Llanvihangel-Ystern-Llewern*, if the name *Monmouth* happens to be the longest in the data set!

3.10.4 Grid references

Store grid references in an appropriate notation for easy transition to a GIS or conversion to an appropriate map projection (e.g. British National Grid references are commonly held as alphanumeric attributes in a single column which require some processing before points can be mapped on a GIS, a more appropriate form of notation would be in two numerical columns e.g. **456344** / **267833** for **SP** 5634467833).

3.10.5 Validation

Get in the habit of ensuring that the data entered into any field in your attribute database makes sense. For example, check that you have not typed the letter 'O' instead of '0' (zero). Another tip is to check that numeric values are within range – for example that a slip of the old typing fingers has not moved your Norman site from 1066 to 2066. It's often helpful to have someone else validate data that you have entered as typos are more easily detected by a fresh pair of eyes. If your data input tools allow you to define validation checks, use them, but remember that – like spelling checkers – they cannot catch all possible input errors.

3.10.6 Numeric data

It is best to use numeric field types rather than text fields if you have numeric data. This can have three benefits. First, confusing characters – such as that familiar O (letter) instead of 0 (zero) problem – cannot be stored in the wrong field type. Second, in many computer-based databases numeric information is stored more efficiently than text and occupies less space. This means that your GIS data set will be leaner and meaner. Third, when data is held in numeric form the data can more readily be manipulated with the arithmetic operators.

If you are using numeric data, also ensure that you use the most appropriate numeric type – integer or floating point. Integer types are used for storing whole numbers and floating point numbers are used for storing numbers which have, or may have, a fractional part.

3.10.7 Data entry control

Where possible the fields should be set up to use dictionaries or thesauri to ensure that typing errors are kept to a minimum and restricted to free text fields, and that terms used to describe real world objects are used accurately and consistently. Adhere to established appropriate project data standards (e.g. the RCHME/English Heritage Urban Archaeological Database Data Standards). If no project standards exist, adhere to the data standards of the digital archive for your data whether that be the SMR or the ADS. Remember that your data will need a home if it is to remain a useful and accessible resource in the future, and it is your responsibility to ensure its compatibility with other data sets of a similar spatial or temporal resolution.

3.10.8 Confidence values

These indicate the level of certainty that is associated with an entry in the attribute database. For example, your certainty that the location, identification, dating, etc. of the object is accurate. It is very good practice to maintain this information at all times.

3.10.9 Consistency

Try to ensure that the codes used to record your attribute data are consistent. Ensuring consistency is especially difficult when data entry is performed by more than one person, or if data entry is carried out incrementally over time. The use of thesauri and documentation standards can be helpful in ensuring consistency within your database and between your database and others. See Appendix 2 for a list of standards that may be appropriate.

3.10.10 Dates

Calendar dates should be recorded in a date field-type rather than character field-type to avoid the loss of crucial data when transferring into different software packages. Be aware some software will not prompt you if you are about to lose data due to incompatible field types.

3.10.11 Documentation

The most important thing of all is to document the way you have organised your database and entered information into it! All of Section 5 is devoted to this topic. It is essential that source-specific information is recorded as and when data is generated, as this task becomes increasingly difficult retrospectively. Where did the source data originate from, what was the scale at which it was prepared, if based on others' work where can this be found, and what are the copyright restrictions involved in its use by a third party? What levels of accuracy were accepted and what errors were recorded during digitization etc? What data standards were adhered to (dated if possible, as revisions will occur) and what naming conventions have been adopted.

Section 4: Structuring, Organising, and Maintaining Information

4.1 LAYERS AND THEMES

The terms *layer* and *theme* are used almost interchangeably by many people – archaeologists and GIS practitioners included – yet are given very distinct meanings by some software suppliers and in some specific disciplines, for example in Computer Aided Design (CAD). For the purposes of this guide these terms will be used as follows. A **theme** is a collection of like objects, for example 'pottery', 'Iron Age sites', etc. A **layer** is a group of specific objects within a *theme* – for example, 'Stamford Ware' within the pottery theme or 'Hillforts' in the Iron Age site theme. In order to avoid confusion, it is important that the names given to such themes and layers are both descriptive and free from ambiguity.

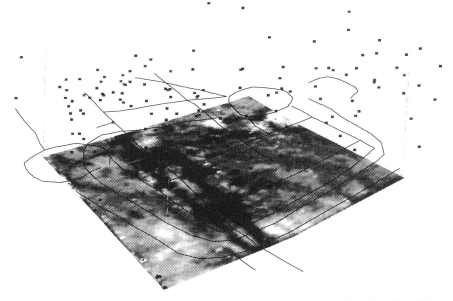

Figure 2: Layers and Themes – Figure created by Peter Halls using data from the Cottam Project directed by Julian Richards. Image copyright © Archaeology Data Service. A layer from the Cottam Project GIS (coin and metal artefact findspots) represents the metal object theme. Two other themes are illustrated in this image, one for aerial photograph interpretations (line data) and another for geophysical survey data (raster image).

The purpose of the theme/layer approach is to provide a framework for collecting together objects of similar nature – in terms of either representation and/or descriptive type. Thus, different Iron Age site types might be gathered together because they are related in terms of both the representational type – a line or point object – and because of their nature or purpose – delineation of the landscape location selected for settlements in the Iron Age. In the same way, a database of finds of pottery might be defined in locational terms as a collection of points, each of which might relate to an individual object, or closely related group of objects.

4.2 CHOICE OF VECTOR, RASTER OR COMBINED FORMS OF SPATIAL DATABASE

The choice of vector, raster, or combined, forms for the spatial database may be determined by the GIS in use. For example, you cannot easily use the vector model within a raster GIS such as GRASS or IDRISI. Similarly, a vector GIS such as pc-Arc/Info cannot manipulate raster data.

Vector means of managing and manipulating the data are to be preferred for handling information relating to discrete points, delimited boundaries, alignment of linear features, etc. Thus a vector model would be used for storing, and manipulating, an excavation plan.

Raster means of managing and manipulating the data are to be preferred for handling continuous information such as altitude (see Digital Elevation Models, below), vegetation, etc., and are the digital form in which information from Geophysical Survey, Aerial Photography, and other forms of Remote Sensing and non-invasive survey, are delivered.

Where both data types are required to be used *together* a GIS capable of manipulating both is required – such as ArcView (with the Spatial Analyst option installed) or Intergraph MGE (with the Grid Analyst option installed).

When combining and integrating information from a variety of sources the following points should be kept in mind:

- All spatial data must be recorded in the same co-ordinate system. Data which are recorded to some other system must be transformed/projected to the required co-ordinate system.
- All spatial data should be to the same *spatial resolution*, or *scale*. It is not possible to get meaningful results from the combination of spatial data recorded to a *scale* of 1:250, as might be the case for an excavation site plan, with road alignments recorded to a scale of 1:250,000. In the former example 1mm represents 25cm, and in the latter example represents 250m. Spatial data recorded to *scales* of greater than around 1:10000 involve considerable generalisation of alignments to avoid features conflicting. This is especially true of *paper* maps drawn to such scales.
- Non-spatial information to be combined, or integrated, must use the same field definitions, encoding regimes, etc. Where different schemes are used it will be necessary to convert or translate the data to the required scheme.

The National Geographic Data Framework (NGDF), recently established in the UK, is producing a series of guidelines for the definition and storage of spatial data such that it has maximum potential for future use. These guidelines include documentation standards. See http://www.ngdf.org.uk/ for more information.

From the point of view of specifically *archaeological* data, the recommendation must be to use one of the data formats defined in Section 6.

4.3 COMBINING AND INTEGRATING ATTRIBUTE DATABASES

4.3.1 Data standards

Successful database integration relies on the implementation of data standards. These aim to facilitate the production of a common frame of reference for archaeologists, endorsed by the profession as a whole and implemented in a widely compatible national network of databases and digital archives.

Currently core data standards are being defined for many fields of archaeology, from portable items such as MDA Archaeological Object Thesaurus (MDA 1998) and the International Guidelines for Museum Object Information, produced by the International Committee for Documentation (CIDOC), to the draft data standards for SMRs and revisions of the RCHME Thesauri of Architectural Types, Monument Types, and Building Materials. Another useful resource is MIDAS (RCHME 1998). Outside of the profession, essential standards have been set for such data sets as British postal addresses (BS7666), and international naming conventions for countries (ISO3166). See Appendix 2 for a list of appropriate data standards.

The basic process involved in the integration of data from external databases relies on compatible field structure. This means that complementary fields in both the source and target databases must be of a compatible type (Integer, Floating Point, Date, a Character field of an appropriate length etc.) to avoid the loss of data during the integration process.

Some features of certain databases (e.g. DBASE memo fields) are difficult to export to other systems and may require specialist advice to avoid their loss. The new data should be date stamped digitally by the computer operator and a record kept of its source and ownership.

4.3.2 Integrating paper records

Data can be extracted from documents and typed manually into an existing database, or whole reports can be captured speedily using a commercially available optical character scanning suite. These convert scanned text into digital characters which can be saved into a variety of word processor formats. The character interpretation is never 100% effective and will require spell-checking and proof-reading before it is used, but this method can save a great deal of time, especially when capturing printed table data. Most often, the integration of paper-records will involve some form of manual input, often involving a number of separate individuals over a considerable period of time. Here the importance of adherence to existing standards and guidelines cannot be over-stressed. Such a process often involves a great number of decisions that directly affect the quality of the source data sets, as often very descriptive information is broken down into the discrete thematic field structure of the database. To ensure that the resultant database is usable it is important to record such decsisions and ensure that a degree of consistency is adopted throughout the process.

4.4 DERIVED DATA

You will often be using data derived from other sources when creating or managing a GIS data set. There are often important considerations in documenting derived data sets, as discussed in Section 5. **When deriving data from another source, or when making use of derived data, it is the responsibility of the data *user* to ensure that any intellectual property rights belonging to the initial data creator(s) are respected.** In some cases this *may* simply be a requirement to acknowledge the originating source, in other cases a royalty payment may be due for some part of the data to be used. Be sure to check out the situation in advance.

4.4.1 The Digital Elevation Model

One of the core components within many GIS databases is the Digital Elevation Model or DEM. This sub-section will look briefly at the principal pathways you can take to create a DEM, special issues that practitioners should be aware of relating to accuracy and integrity, and the specialised process-related metadata that should be recorded.

DEM vs DTM. There is considerable confusion between these two terms – which many people use interchangeably. Some people also use the term Digital Surface Model (DSM). DSM appears to be a synomym for DEM, but with the added possibility of being a component of a stack of surface models, for example modelling atmospheric or subsurface layers. DSMs are not yet in common usage.

Strictly speaking, the term DTM, Digital Terrain Model, should be reserved for those models of reality which include information relating to surface texture, etc., in addition to information regarding elevation. The term Digital Elevation Model, DEM, should be reserved for representations of altitude alone.

There are few genuine DTMs around yet – the concept is established but the tools to display, or visualise, information of this nature are not yet fully/widely available.

A DEM normally consists of a regular matrix of elevation values, from which altitude functions such as slope and aspect can be calculated, and which may be rendered for visualisation as isolines (contours), perspective or panoramic views, etc. A DEM is typically described in terms of its horizontal resolution. Resolution, for a DEM, defines the horizontal and vertical precision by which the information is recorded. A typical example might define the horizontal grid to be fifty metres – by which it is meant that the information in the DEM is arranged with one value every fifty metres in each of the co-ordinate directions. The resolution of the vertical element of the information will indicate whether the value represents the computed average elevation for that, say, fifty metre cell, of the elevation at the mid point as well as defining the range of accuracy to be expected of the elevation values, possibly plus or minus two or three metres.

Sometimes the surface elevation information does not form a regular matrix, but rather comprises a collection of measured locations with altitude. A DEM constructed from such data may be *interpolated* to form a regular matrix, or the surface may be represented by linking the measured points within a Triangulated Irregular Network (TIN). The TIN structure, of triangular facets each with a measureable slope and direction, is an efficient storage mechanism from which a regular matrix can readily be derived when necessary.

TINs have another useful property: they can be *stacked*. This means that they can be used to represent *layers* of information, for example atmospheric layers, subsurface archaeological or geological stratigraphy, etc.

Resolution, in the context of a TIN, is a simpler concept than for DEMs formed from a regular matrix of values. The vertical resolution remains the same but the horizontal resolution is a function of the precision of the co-ordinates defining the data points and the number and distribution of the data points with respect to the surface morphology. A primarily flat surface will require fewer points than a rugged or undulating surface.

Contour lines are not a good form in which to store elevation data. Contours are derived data, data interpolated from information of altitude at known points, and in themselves offer no information about the surface morphology between them. Contours *may* be an effective way of illustrating the third dimension (altitude or depth) on two dimensional paper, but are a poor technology for storing altitude information that may be used analytically.

A DEM held as a regular matrix suffers the same disadvantages due to size as does the raster data model. Although the co-ordinates of each cell in the data set can be derived from the co-ordinates of the origin and from the number of cells in each direction and their separation, a value must still be stored for every cell. The computer file size for such a data set is thus the product of the number of rows and the number of columns. For example, an OS LandForm Panorama *tile*, 20km square and with a 50m cell separation, holds 401 rows and columns – 160,801 cells. The same area in the LandForm Profile data set, with a cell separation of 10m, comprises 402,002,500 cells. In computer storage terms, the coarser data set would require around 629Kbytes; the finer resolution data set requires some 1.57Gbytes (note difference in units). If the elevation data is held in floating point form (fractional numbers) then the storage requirements *may* be doubled, depending upon the individual computer system. It is less easy to give guidelines on the space requirements of TIN data sets ... a *rough* guide might be (number of points times 6) times 4 bytes. Thus a TIN with 900 data points would require 21600 bytes (21Kbytes). This figure may vary according to the implementation of the TIN structures.

There are several potential sources of elevation data:

- Primary Survey. EDM, theodolite, etc., measurements. These data are typically irregular in their ground coverage and are thus ideal for TIN data structures
- Photogrammetry. Conventional photogrammetry using stereo images, generates isolines, contours. These must be converted into digital elevation data, typically via the TIN data structures.
  Digital photogrammetry uses stereo digital images and typically produces a regular matrix of values, a DEM
- Synthetic Aperture Radar (SAR) altitude measurements. Satellite radar imagery which generates a regular matrix of altitude values directly, a DEM

Several times in the above discussion the terms *interpolation* or *interpolated* have been used. In the context of DEMs these refer to the technique used to approximate the altitude of points for which there is no measured data. The purpose of interpolation is to attempt to regain a representation of the actual surface morphology.

There are perhaps three techniques commonly used in GIS to perform such a task: each has specific capabilities dependent upon the nature of the data. These three techniques are *Linear interpolation*, which effectively runs a straight line between the points with altitude values, the

Cubic Spline, which interpolates a smooth curve through the given data points, and *Statistical interpolation* using *Kriging* semi-variograms. For further discussion see Chapter 8 of Burrough (1986).

4.5 COPYRIGHT ISSUES – AN EXAMPLE FROM THE ORDNANCE SURVEY

The following notes were based upon OS information current at the time of writing and are intended as a guide. The most recent OS documents should always be consulted for the definitive conditions.

All derived data that you may use in your GIS will be copyright by someone whether this is you, your organisation, or someone else entirely. It's very important to keep careful track of who owns the copyright on each piece of information you incorporate into your GIS data set as this will affect how you can publish your data and who else can use it. An example of the complexities of copyright comes from the Ordnance Survey of Great Britain.

4.5.1 Requesting Permission

The Ordnance Survey (OS) requests that users of their information ask permission before any procedure requiring copyright clearance is undertaken. As OS data is copyright by the Crown, especially rigorous regulations are applicable which makes it particularly important to understand their copyright requirements **before** undertaking a GIS analysis.

The Copyright, Designs and Patents Act 1988 (HMSO 1988) defines Crown copyright and states that it would be infringed if any person or organisation reproduced a copy of Crown copyright information without first having permission. Copyright is infringed when:

- the copying is by hand or by mechanical means
- the copying is direct or through a fresh drawing in whole or in part, or from a map or document based on Ordnance Survey material.

One exception to this need for permission is a 'fair dealing' clause, which generally means that you do not need permission to make up to four A4 size copies of a map that you own for private study, research, criticism or review. The OS define 'educational purposes' in terms of paper map extracts for the purposes of teaching map reading and interpretation, etc., and for related assessment of a student's progress. Research usage is limited to 'university rersearch projects only'. This generally excludes research supported by 'external' sponsors including higher education funding councils.

Another exception is the use of OS material in connection with any proceedings on parliamentary, judicial, Royal Commission, or statutory inquiry matters. In these cases OS material may be used without permission but cannot be used in a publication (Ordnance Survey 1996).

4.5.2 Citation

Whether or not you need advance permission to use OS material, you **must** cite the fact that you are using their information.

With OS material that is out of copyright, you **must** include the statement 'Reproduced from the (year of publication) Ordnance Survey map' (Ordnance Survey 1996).

4.5.3 Are all OS maps copyright?

Strictly speaking, **all** legitimate use of OS material will carry with it an appropriate citation together with the phrase 'Crown Copyright Reserved'. The OS encourages you to seek confirmation from them directly whether or not material which does not bear this inscription is subject to their copyright.

4.5.4 Publishing

The OS define 'publishing' as 'selling or giving away any publication which contains our material. By publication, we mean sheet maps, books, journals, brochures, leaflets, catalogues, and so on' (Ordnance Survey 1996).

Generally royalties are not charged on publications of academic research that are not intended to make a profit, but you will need to get OS permission before publishing and you will need to acknowledge OS copyright in the publication. There are various scales of royalty charges for publishers of things other than academic research and discounts are offered for registered charities, etc. The scale of royalties depend in part upon both the proportion of OS material in the product and the OS perceived 'usefulness' of the final product. OS Copyright Leaflet 4, dated April 1996, defines all these provisions.

Generally the OS do not allow any of their material to be published on the World Wide Web (Ordnance Survey 1996).

4.5.5 The National Grid

Believe it or not, the UK national grid is even covered by Crown Copyright on Ordnance Survey maps. Generally you can copy the national grid without having to pay a charge, but you must print the following acknowledgement:

> "The grid on this map is the National Grid taken from the Ordnance Survey map with the permission of the Controller of Her Majesty's Stationery Office." (Ordnance Survey 1996).

It's not the national grid, *per se*, that is copyright. The national grid is a standard Transverse Mercator map projection, using the Airey Spheroid and a false origin to the south west of the Isles of Scilly. Since it is a mathematical transformation, determined by people long since dead, it is not of itself copyrightable.

However, the OS **usage** of the national grid – displaying grid co-ordinates with two letter codes for the 100km squares – **is** Crown Copyright. The OS claim 'ownership' of any national grid reference (NGR) which is given in the 'NY 123 456' form. It is our understanding that the OS have been successful in obtaining injunctions against commercial use of this form where no royalties were paid to the OS.

This means that the example reference, NY 123 456, given above **is** OS Copyright whilst its numeric form, 312300 445600, isn't.

4.5.6 Use of OS Data (paper maps and digital data)

The general case is that everything conveyed within an OS product is Crown Copyright. Clearly, the OS cannot claim copyright of the name of a settlement, for example, but they can claim ownership of anything relating to their survey effort that delineates the extent of that settlement. Also, equally clearly, any errors in their data are OS Copyright. This may seem strange, but error matching is one of the mechanisms that the OS uses to determine whether or not someone has violated their copyright.

4.5.7 Ground control points and OS maps

Taking ground control points from OS maps is a restricted right under Copyright Law. Restricted rights are not covered under the standard licence granted by Ordnance Survey, and instead are only granted on a case by case basis. A charge is almost always made.

Using OS data (maps, digital data) to specify position of other data (maps, images, etc.) gives the OS an intellectual property right in the resultant data, map, etc. Indeed, the OS claim that they **own** the positional content of the new data. It is thus an offence to pass this information on to a third party without the explicit, and prior, permission of the OS. The OS will seek a royalty for any information which they consider to be a commercial product, or to have commercial potential. 'You risk breaking Crown Copyright if you try to capture your own data which has been fixed from Ordnance Survey material.' (Ordnance Survey 1996).

Of course, if the data are surveyed in from a GPS point, for example, and the co-ordinates are specified in full numeric form (e.g. 312300 445600 from the above example) the OS are not involved. The numeric form is far more useful in relation to digital data anyway, so this is a good habit to get into! Differential GPS, with a potential precision from less than a cm up to a metre or two is readily available now. It can be hired cheaply, is already in wide use, and thus may already be accessible to you. It's a good idea to keep accurate field records about GPS use in order to prove your exemption from OS claims.

4.5.8 Digitising OS Maps

The OS do not generally grant permission to digitise any OS copyright material for which there is an existing digital equivalent. Now that the whole OS production system is digital, that means that you will need to argue a very specific case to get permission. Wanting to digitise a map at a scale of 1:25000 when an OS digital map is available at 1:1250, will probably not gain you permission. When the OS do grant digitising permission there is generally a data capture royalty to be paid.

4.5.9 The worst case scenario

Copyright law is a serious matter, and it is best to seek the proper permissions in advance. According to the OS:

> "If you reproduce our maps or use our digital data without our permission, this is known legally as an infringement of the Copyright, Designs and Patents Act 1988. If you do this

you are stealing the results of our work and we will take legal action. We will decide if you must pay damages and if you must destroy the copies. Our minimum damages will be the same as the royalty fees which apply at the time we discover the infringement, plus 25%." (Ordnance Survey 1996).

4.5.10 Other issues in using data derived from the OS

Whilst the OS claim their data to be complete they do not claim 100% positional accuracy. Indeed, there is a residual error which relates to the early County Series to National Grid transition which the OS currently estimate needs an investment of over 40m pounds to resolve. Their present approach to this problem is to ask customers to notify them of corrections so that they get a piece-meal update ... for free! City centres are least affected by these errors.

The biggest issue is that of OS claiming intellectual property rights in 'knowing where you are'. This means that much existing data, whilst free from liability as long as it remains solely with its collector, may become liable to copyright royalties the moment it passes to another person. It also implies that the use of GPS technology for 'fixing' position for **all** future field data collection work should be seriously costed – by this means the collected data would become free of OS claims.

4.5.11 Would you like more information?

Additional useful information can be found on the OS Web pages. For copyright see the http://www.ordsvy.gov.uk/about_us/copyrite/index.html and the pages that follow.

Section 5: Documenting your GIS Data set

5.1 *WHY* DOCUMENT YOUR DATA?

Working with your Geographic Information System on a regular basis as you do, you probably have a pretty good idea about what it contains, the area of the country it covers, and what its major strengths and weaknesses are likely to be. You know, for example, that your data cover the city of York, that period information is only stored to the nearest century, and that the aerial photographic interpretation to the south-west of the city is a bit dubious.

5.1.1 Documentation for others

Data offered to the ADS, however, may potentially be used by researchers from many different parts of the planet, and with widely varied levels of expertise. *They* have no way of knowing anything at all about your data unless you tell them.

In order to make sure that the maximum amount of information is delivered to the user whilst involving you, the depositor, in minimal effort, the Archaeology Data Service has developed a number of procedures to standardise and simplify the documentation process.

5.1.2 Documentation for you

Some form of record about your data – and about what you've done to it – is also, of course, undoubtedly useful within your own organisation. Even using data every day, it is still possible to forget about where *some* of it came from, or how the data you currently used were originally compiled from various sources.

This guide introduces the issues relevant to both types of documentation, as well as discussing the detail relevant to one or the other.

5.2 LEVELS OF DOCUMENTATION

In documenting something so complicated as a Geographic Information System, it is possible to enter into great detail, and record everything from the data sets comprising the GIS to the sequence in which individual commands were applied to the data in order to produce your current system.

As with most things, there are situations in which great detail is required, and others where a more slimmed–down level of recording might be more appropriate. It is generally up to you as creator, maintainer, and primary user of your data to decide how much documentation is justified, and to select a suitable level for the language of your documentation; is 'cleaned coverage' appropriate, for example, or is your situation such that the more expressive:

> 'Using the Arc/Info `clean` command, tidied up the new pottery layer in order to remove errors introduced whilst digitising from the paper map. The command was
>
> `clean pottery # # # poly`'

is more suitable? The former has implications for decreasing the interpretability of your documentation, whilst the latter has implications for the effort required in producing this level of detail.

In documenting data which are to be made available to others, it is often necessary to describe things more clearly – and with a greater degreee of contextualisation – than is normally the case for internal use.

5.2.1 'Documentation' versus 'Metadata'

This guide, and many other documents distributed by the Archaeology Data Service, talk at length about a concept labelled as *metadata*. The term itself is a piece of jargon, but it serves as a useful label which is increasingly understood both within archaeology and in the wide world outside our discipline.

Several definitions are offered for metadata, but one which might usefully be given here is that metadata is the means by which your *data* are transformed into *information*, interpretable to and re-usable by those other than yourself. In other words, metadata is a label for the extra details associated with any data set which enable someone else to place them into some form of context. Metadata might include information on the computer format in which the data are stored, the area of the country they relate to, etc.

Metadata in its widest sense may be considered *all* of the documentation conceivably associated with GIS data, but the ADS simplifies things somewhat by using the metadata label only to apply to metadata used for *resource discovery*. As such, information suitable for entry into the ADS catalogue itself, and which can be used to facilitate the discovery of your data by others, can be thought of as metadata, whilst the information you provide which helps people to *use* your data after they have accessed it may be thought of as ancillary documentation.

5.2.2 So... how much is enough?

Well, it depends... If you are documenting data for your own internal use, you are of course free to use as much or as little of what is recommended here as you like.

If, however, you are preparing data for deposit with the ADS, then you will need to comply with the guidelines in section 5.4 on Dublin Core metadata, as well as supplying sufficient documentation (discussed in section 5.3) to enable re–use of your data. Where your data are particularly complicated, ADS staff may recommend other – specialised – documentation to accompany them, and will hope to enter into dialogue with you at an early stage in order to define this.

At any stage during the documentation process, ADS staff are available to help in clarifying these guidelines, and to aid in interpreting their generality in the context of your specific requirements and data.

5.3 INFORMATION TO BE RECORDED

It is generally a good idea to start recording information about your data as early as possible, and ideally you should begin recording as soon as you start using or creating the data. If you wait until just before depositing with ADS to start creating metadata and documentation, it will be difficult for you to provide some pieces of information at all, and far harder to write most of the rest than it would have been at the time you were actually *doing* it.

Assuming that you choose to record relevant details as you go along, it might be useful for you to start a formal **log book** of some kind. This way, it will be easier to find information later, rather than having to rifle through various old envelopes, scrap paper, and whatever else you scribbled on at the time.

Within this log book, it is normal to record such general details as the software you are using, the versions thereof, and the type of computer (PC, Macintosh, Sun workstation, etc. rather than Dell, Viglen, Compaq, etc.) you are running it on. As time passes, people discover problems with earlier versions of software, and if someone finds out that *SuperGIS* version 23.7 displaced all green lines on maps by 3mm, then it is undoubtedly useful for you to be able to look back through the log book and find that all your maps displaying public rights of way were created three years ago using *SuperGIS* 23.7. Knowing there is a problem, you can do something about retrospectively fixing it with adequate documentation.

5.3.1 Sources of data

Information about where the data you use are acquired from is one of the most important things you can record whilst constructing and using a GIS.

Data are acquired from numerous sources, including Ordnance Survey and other mapping agencies, local authorities, special interest groups, etc., and are gathered and displayed at a wide variety of – often different – scales or resolutions.

Each of these sources are of value for a different set of purposes, and each brings with it a different set of problems; data acquired at 1:50,000 scale, for example, may be ideally suited for plotting maps of artefact distributions, but wholly improper for recording the layout of individual excavation trenches (1 centimetre on a 1:50,000 map, after all, is equivalent to 50,000 centimetres, or 500 metres, on the ground).

In order to aid the user in deciding how best to incorporate your data within their own work, it is desirable to provide them with information such as the scale or resolution of the original survey, scale or resolution at which that survey was digitised into the computer, assumed errors from the data capture process (often expressed as a Root Mean Square, or RMS, error on printed maps), and the method by which the data were originally acquired (although both ultimately plotted at a scale of 1:100, a user will presumably be interested to know that one topographic data set was constructed by survey with measuring tapes and dumpy level, whilst the other is the result of a detailed survey by state of the art Total Station Theodolite).

Ownership of data is also an important attribute to record about any data set, and may well prove quite complex. Data owned by the Ordnance Survey, for example, might be used by North Yorkshire County Council to derive a new data set, 'owned' by the County Council. This, in turn, is used by York Archaeological Trust to derive a new data set, now 'owned' by them. Although little, if any, of the original Ordnance Survey resource may survive in this latest incarnation of the data, Ordnance Survey in reality continue to hold intellectual property rights which should be recognised and which may well affect the ease with which, for example, York Archaeological Trust could later *legally* sell 'their' data to Yorkshire Water.

Complicated *data trails* such as this are extremely common with digital data, and it makes life easier for everyone if the evolution of every data set is tracked through every reincarnation.

In short, then, a *non-exhaustive* list of the information you might wish to record during your everyday creation, collection, and use of data includes:

- Computer hardware used
- Computer software used
- Date the data were captured/purchased/whatever
- Who did the work
- Data source ('bought from Ordnance Survey', etc.)
- Scale/resolution of data capture
- Scale/resolution at which data are currently stored
- Root Mean Square error or other assessments of data quality
- Purpose of data set creation, where known
- Method of original data capture (Total Station Survey, etc.)
- Purpose for which *you* acquired the data (might differ from the previous information where the data were *created* by someone else for one purpose, and bought from them by you for another)
- Complete history of data ownership/rights.

5.3.2 Processes applied

As well as recording information such as that suggested above, most of which will probably only need recording once when you start work with a data set, it is also extremely valuable to log the manner in which data are manipulated and modified. Not only does this allow you to keep track of – and back-track *from*, if necessary – changes you make to the data, but it also allows you and others to work out how data you lifted from your local Sites & Monuments Record, for example, and incorporated into your own GIS differs from those same records still residing in the SMR. How many records have you enhanced? For how many have you had to re-enter the grid references, as you discovered that those provided by the SMR actually placed sites in the North Sea?

The sorts of information you may wish to consider logging for these purposes include:

- The date of any change/modification
- The reason for any change/modification
- The record numbers affected by the change
- Relationships to other resources; where, for example, you derive a new GIS data set by passing a mathematical filter or some other modification through an existing data set, you may wish to record the relationship formally between the original data and the new set.

Where you edit an existing data set to correct spelling in text fields, or some similar operation, it makes more sense to simply record this as 'Corrected spelling throughout data set' and give the numbers of those records altered if relevant, rather than to list every single correction made to every single record. For processes such as converting an elevation matrix to a Triangulated Irregular Network (TIN) or an equally drastic data set-wide modification, it is worth recording the parameters you used in undertaking this process so that you – and others – may repeat or undo it in the future.

5.4 DUBLIN CORE METADATA

5.4.1 What it is...

Much of the information recommended for you to record is most useful when it comes to actually *using* your data, and as such will probably only be downloaded by a potential user at the same time as they access your data. Certain pieces of information, though, are key in aiding the user in *finding* your data in the first place, and it is these that are explored in this section.

The Archaeology Data Service, along with a growing number of organisations around the world, advocates use of the *Dublin Core* for recording the information that helps potential users to find – and simply evaluate – your data. This information is known as 'resource discovery metadata'; information about your data (the 'resource') that helps people discover it.

Through more than three years of international development, the Dublin Core has evolved to become a series of fifteen broad categories, or elements. Each of these elements is *optional*, may be *repeated* as many times as required, and may be *refined* through the use of a developing set of sub-elements. The use of the Dublin Core within the Archaeology Data Service is discussed further elsewhere (Miller and Greenstein 1997, Wise and Miller 1997), and the current element definitions laid down across the Dublin Core community are available on the web.

The fifteen elements of the Dublin Core may be simply defined as:

- **Title**
  The name given to this resource by its creator. The name need not necessarily be unique.
- **Creator**
  Those persons or organisations responsible for creation of the resource being described, its source or surrogates, whose involvement is considered worthy of inclusion for the purpose of discovering said resource.
- **Subject**
  That which the resource is about, preferably using terms drawn from a controlled vocabulary such as the *Thesaurus of Monument Types*.
- **Description**
  A description of the content of the resource, such as a book abstract or a synopsis of database contents.
- **Publisher**
  The entity(s) responsible for facilitating availability of the resource, such as a publisher, distributor, or a corporate entity.
- **Contributors**
  This element is not used by ADS, as it causes too much confusion amongst users.
- **Date**
  Dates associated with the creation and dissemination of the resource. These dates should

not be confused with those related to the *content* of a resource (AD 43, in a database of artefacts from the Roman Conquest of southern Britain) or its *subject* (1812, in relation to Tchaikovsky's eponymous overture), both of which are dealt with in other elements.

- **Type**
  The general form of a resource, such as text, image, etc.
- **Format**
  The format, either physical (e.g. book, CD-ROM) or electronic (e.g. DXF file, HTML web page), of the resource being described.
- **Identifier**
  A text string or number used as a unique identity for the resource. Examples include a resource-specific URL or ftp address, an excavation site code, or archival museum shelving number.
- **Source**
  Any important earlier work(s) from which this resource is derived.
- **Language**
  Language(s) of the intellectual content of the resource (i.e. English or Latin, rather than C++ or Pascal).
- **Relation**
  Relationship to other resources. For example, the GIS resource being described might be part of an SMR or landscape study, both of which might have entries of their own to which this element could provide a link.
- **Coverage**
  The spatial and temporal extent(s) pertaining to the resource. In both cases, coverage relates to the *content* of the resource, rather than to its collection or management. Likely coverages include the spatial location (whether a grid reference, place name [Skara Brae], or more ephemeral locator) and temporal period (whether a date, date range, or period label [Neolithic]) of the Skara Brae village and exclude the location of the museum in which the artefacts might now be found.
- **Rights**
  This element is intended to be a link to a copyright notice, a rights management statement, or a software tool capable of providing such information in a dynamic fashion. The intent of specifying this field is to allow data providers a means to associate terms and conditions or copyright statements with a resource or collection of resources.

5.4.2 ...and how to create it

The Archaeology Data Service is working with other organisations in order to create a number of tools that will make it easier both for you to *create* this information, and for us to *read* it when you send it to us. This work is ongoing, and you should contact us before creating records for deposit with ADS in order to find out the latest developments in this work.

Until such time as these tools are available, it will be necessary for you to create these Dublin Core records manually, and send them through to us either on a computer disk or via e-mail. These options are discussed in a little more detail in Section 6.

With complex collections of computer files such as those present in most archaeological GIS, it is extremely difficult to draw up simple rules defining what you should create metadata records *for* (the GIS as a whole, every 'layer', every original data source, etc.). As a basic guide, it is sensible for you to create one record describing the GIS as a whole, plus one subsidiary record for each major resource 'type' stored in the system. If, for example, you created a GIS for Hampshire which recorded Neolithic burial monuments and Roman settlement patterns (well – you *might*!), it would seem sensible to create one record for the whole, one for the Neolithic part, and one for the Roman part, giving three records of which one (the whole GIS) is the 'parent' and two are 'children'. If in doubt, contact ADS for advice.

An important factor in ensuring that a user can sensibly compare records you create with those provided by others is the use of standardised terminologies and modes of expression. Whilst certainly *not* wanting to restrict all archaeologists to the use of a single standard for recording, or a single thesaurus for controlling terminology, the ADS does believe in the use of standards in general. As utilised by the ADS, the Dublin Core system allows users to identify a 'SCHEME' which controls the terms stored in any one occurrence of a Dublin Core element. Thus, a user could identify the *Thesaurus of Monument Types* (RCHME 1995) as a SCHEME for the Dublin Core Subject element, and describe their resource using terms drawn from this thesaurus. As all of the Dublin Core elements are repeatable, the user could then – if they wanted to – repeat the Subject element and define the Getty's *Art & Architecture Thesaurus* as their SCHEME. Importantly, each use of a Dublin Core element should only include terms drawn from *one* SCHEME. Where absolutely necessary, it is possible to enter information as 'free text', not qualified by any SCHEME, but such use of free text makes it far harder for users to search across resources meaningfully, and should thus be avoided. A non-exclusive list of relevant SCHEMEs for providing archaeologically relevant controlled terminology is discussed elsewhere (Wise and Miller 1997), and ADS staff are available to offer advice or clarification when needed.

5.5 ANCILLARY DOCUMENTATION: WHAT TO SUPPLY AND WHY

Possibly the single most important piece of information you can provide above and beyond the Dublin Core catalogue entries discussed above is an idea of your *data model*.

This model enables potential users to discover relatively quickly what sorts of information your GIS will probably hold, and allows them to work out how the whole thing is tied together.

For a typical archaeological GIS, the information that might usefully be represented in a data model submitted to ADS includes:

- a list of field names (and definitions) for your database e.g. **Address**: The postal address of the archaeological intervention being described.
- a diagram depicting the relationships between database tables, if relevant.
- a list of map/coverage/'layer' names (and definitions) e.g. **modernyork**: The modern streetplan for the study area, extracted from Ordnance Survey 1:1,250 scale digital mapping.

Other than the data model itself, much of the information this section advocates for entry into your project log book can usefully be passed on to the ADS in digital form (see Section 6), as it is equally useful to *others* trying to make use of your data as it was to you.

Section 6: Depositing Information –
Archiving your Data set

6.1 THE ARCHAEOLOGY DATA SERVICE – YOUR FIRST PORT OF CALL

The Archaeology Data Service collects, catalogues, manages, preserves, and encourages re-use of digital resources created by archaeologists. It is one of five service providers in the Arts and Humanities Data Service which provides archival services for a range of disciplines including history, text studies (literature and linguistics), performing arts, and visual arts. Archaeologists who are creating GIS-based resources are invited to consider archiving these resources with the ADS.

The ADS archives and catalogues high quality digital resources of long-term interest to archaeologists. The ADS provides an archival home for any archaeological data of interest to UK archaeologists. The scope of the ADS collections is thus international, although priority is generally given to the archaeology of the British Isles. Where existing archival bodies work to preserve digital collections, the ADS will collaborate with these bodies to facilitate more uniform access to on-line information. The ADS Collections Policy provides more detailed information.

Depositors are asked to specify that the ADS may have a non-exclusive license to distribute their data sets. In the interest of archaeological research, the ADS encourages depositors to make their data available to the broadest possible spectrum of archaeologists. It is possible, however, to allow access only to those who will use your data for educational purposes. Potential users are asked to sign an user agreement in which they detail how they will use data obtained via the ADS.

For the most up-to-date information about archiving data, please refer to the ADS Guidelines for Depositors.

If your data are not appropriate for deposit with the ADS, address enquiries to the AHDS Service Provider serving the discipline area in which you are currently working. If you are in doubt, please refer to the AHDS Collections Policy or contact the AHDS Executive for additional information.

- The History Data Service
- The Oxford Text Archive
- The Performing Arts Data Service
- The Visual Arts Data Service

6.2 FOUR REASONS FOR DEPOSITING GIS-BASED DATA WITH THE ADS

6.2.1 Ensuring preservation

Archaeology is in a special position in that much data creation results in the destruction of primary evidence. Increasingly, the digital record may be the only record of precious research materials. With the ever-increasing pace of change in computer hardware and software, in a few years' time that data may be lost forever. The best strategy for long-term preservation of archaeological data is for them to be systematically collected, maintained, and made accessible to users operating in very different computing environments. Data deposited with the ADS will be migrated through changing technology so their intellectual content will be available in the future. Resources in the ADS collection will also be professionally catalogued according to appropriate standards. Information about them will be made accessible through an online catalogue, and the conditions under which data may be distributed to users can be specified by the depositor.

Digital preservation differs from traditional archiving. Archives traditionally preserve physical objects (e.g. paper, photographs, microfilm) which carry information. Digital archiving is about preserving information regardless of the medium on which that information is stored. The content, structure, and context of the information must be preserved in order for digital records to be migrated from one medium and format to the next. This is because diskettes and other magnetic media degrade and software and hardware change rapidly: the physical medium on which digital data is stored is impermanent.

Digital archiving strategies do not, and should not, rely on the preservation of a single diskette, tape, or CD-ROM. The essence of digital archiving lies in four main activities: secure backing-up, data refreshment, data migration, and documentation.

Most people have heard of *backing-up*: the act of making duplicate copies of digital data and storing these copies in a secure environment. The creation of secure back-up copies does not adequately protect digital data from the degradation of the media on which they are stored. Three additional steps are required for successful digital archiving.

Data *refreshment* is the act of copying information from one medium to the next as the original medium nears the end of its reliable lifespan. Research into the lifespan of both magnetic and optical media has been conducted. The overwhelming conclusion from this research is that even though magnetic media can be safe for 5–10 years and optical media may survive as long as 30, technology changes much more quickly. Digital media are far more likely to become unreadable as a result of changing technology than they are through media degradation. For example, 10 years ago many archaeologists collected information on 3 inch Amstrad diskettes. These diskettes are completely unreadable by PC machines, and can only be accessed on networked Amstrads – very rare indeed in 1997. If archaeologists had refreshed their data from 3 inch Amstrad diskettes to 5.25 inch diskettes and then to 3.5 inch diskettes, that digital data would still be accessible and safe.

Data *migration* is even more important than data refreshment. Migration is the act of copying digital information from one format or structure into another. One example is copying old flat-field database files into a newer relational database. The functionality of a flat-field database can be maintained in a relational database structure, but without migration the improvements in data handling and retrieval which have led to the widespread acceptance of relational databases would not be drawn upon. Another example is the migration of data between different Computer-Aided Design (CAD) packages. Even though CAD packages allow data to be exported in a 'standard' format called DXF, each programme in fact creates DXF files differently. The unwary can find that without careful migration much of the original information is lost when creating a new file. In order for a digital archivist to migrate digital information successfully, it is necessary to understand the structure of the data fully, and how different parts relate to one another.

Data migration thus relies on the fourth activity mentioned: *documentation*. No digital archivist can successfully preserve data that are not fully documented, because at every step of data migration information can be lost. This leaves archivists with two options: migrating data from one format and then double-checking each entry manually or requiring thorough documentation of the data at the time of archiving so migrations can be carefully planned and tested in advance.

6.2.2 Providing access

Letting others have access to your data is important for two reasons. First, it facilitates communication within archaeology. Second, sharing your data actually helps in its preservation. The more formats a data set is copied into, the greater its chance of surviving.

6.2.3 Professional recognition

By evaluating data offered for deposit, the ADS ensures that its collection consists of high quality and well-documented data sets. This provides depositors with tangible evidence of the quality of their work, and leads to recognition of the effort involved in the creation of digital data. Data sets deposited with the ADS may be promoted to potential users through publicity, workshops, and the online catalogue. Depositors will also receive recognition and citation for their data sets as they are incorporated into future archaeological research and teaching.

6.2.4 Meeting funding agency requirements

If you have received a research grant from the British Academy, the Carnegie Trust, the Council for British Archaeology (CBA), the Economic and Social Research Council (ESRC), the Leverhulme Trust, the Natural Environment Research Council (NERC), or the Wellcome Trust's History of Medicine Programme you are either required or recommended to offer relevant data for deposit with the ADS and AHDS.

6.3 DEPOSITING INFORMATION

6.3.1 How do I deposit a resource?

ADS staff are available to consult with potential depositors by e-mail or telephone during office hours (9–5 Monday to Friday). More detailed Guidelines for Depositors are also available, and this is where you will find the most current information about depositing data with the ADS.

Data sets offered for deposit should be accompanied by:

____ Signed forms which stipulate how and by whom the data can be reused, and how the data and its documentation will be transferred to the Archaeology Data Service.
1. Deposit Licence (to be signed and returned)
2. Access Agreement (for your information)
3. Data and Documentation Transfer Forum (to be signed and returned)

____ Appropriate documentation
____ Appropriate metadata records

Data sets offered for deposit are reviewed by the ADS Management Committee to ensure they are appropriate for deposit with the ADS. See the Guidelines for Depositors for more information about evaluation criteria.

Once accepted, data sets will be scheduled for accessioning. Accessioning procedures include data validation, cataloguing, and mounting. The complexity of the data set determines how long accessioning takes, but it is our goal to complete this process in an average of 3 months. If there is a special reason why data sets need to be accessioned more quickly, please contact the ADS in advance to discuss your requirements.

6.3.2 What detailed information needs to accompany data sets to ensure that they can be re-used by others?

As discussed in Section 5, the documentation which accompanies a data set should enable a third party to make sense of the data. In addition to the documentation suggested in Section 5, here is a checklist of more general information that you may wish to include:

____ Project Title
____ History of the Originating Project
• the purpose of the project
• topic(s) of research
• geographic and temporal limits
• other relevant information

____ Information about Methods
• methods used to create the data set
• methods used to georeference data
• consistency checks
• error corrections
• sampling strategies employed
• other relevant information

____ Details of source materials used to create the data set
• archives interrogated for desktop assessments
• maps used to georeference site grids or surveys

- previous excavations/evaluations of the site
- data selection or sampling procedures
- procedures for updating, combining, or enhancing source data
- description of any known copyrights held on source material

____ Content and structure of data set
- list of filenames and description of contents
- description of identification numbers assigned
- list of codes used, and what they mean
- description of any known errors
- indications of any known areas of weakness
- details of derived variables or coverages
- data dictionaries, if available
- documentation of record conversion to new systems and formats
- description of the record-keeping system used to document the data set
- names of primary project staff
- history of format changes to data set
- history of how the data set has been used
- other relevant information

____ Details of how the data set relates to other archives and publications
- bibliographic references to any publications about the site or project
- information about any archives, museums, SMRs, NMRs, etc. which hold material related to the data set
- information about any non-public material relating to the data set

6.3.3 Deposit formats

The formats that are safest for digital preservation vary with the type of information contained within a file. In this section, recommendations are given for formatting of GIS files, databases, images, documentation, and metadata.

6.3.3.1 GIS files

The ADS is able to accept most major file formats, though Arc/Info formats are preferred as this is the software which we actually used to manage and migrate GIS data sets. We can accept anything from a combination of DXF and DBF files to the following GIS formats:

- Arc/Info export
- Arc/Info ungen
- ArcView
- IDRISI
- GRASS
- MIF/MID
- NTF
- SDTF

- MOSS
- VPF

6.3.3.2 Database files

If you have external databases connected to your GIS system, for example a database containing your attribute data, then you may want to archive these as well. The ADS prefers ASCII delimited files (providing that you tell us how you have delimited them... whether you have used commas, semi-colons or more creative things) and DBF files. We can also accept files in Access and Paradox formats, but *please* make certain that you have not formatted the databases in a way that are specific to these programmes.

6.3.3.3 Image files

It is **NOT** necessary to archive images of every single coverage in your GIS, nor is is necessary to archive images showing all of the ways you used the GIS to play with that data. Occasionally an image may have proven useful to you in a research project and, in order to document the research that you did, archiving that image might be worth more than 1,000 words of documentation. One example is an image showing lithic flakes scattered across a house floor in a pattern that you argued demonstrates lithic production was taking place on site – that single image might be well worth including.

We prefer images to be in TIFF format, but can accept BIL, BMP, CGM, Geotiff, GIF, JPEG, PhotoCD, and PNG files too.

6.3.3.4 Documentation to accompany your GIS, database, or image files

Your data set – the GIS files, database files, and image files – will need to be accompanied by detailed documentation as described in Section 5 and earlier in Section 6. We prefer this information as either ASCII text files or RTF files, but can also accept HTML, LaTeX, ODA, PDF, Postscript, SGML, TeX, Word, or WordPerfect files.

6.3.3.5 Metadata to accompany your GIS, database, or image files

As mentioned in Section 5, tools are being developed to assist in the automatic creation of metadata records. These tools will take the form of interactive forms on the Internet, and assistance in downloading metadata information directly from existing databases. Information about these tools will appear in the ADS Guidelines for Depositors. In the meantime, though, life must go on. We prefer metadata to be sent in either ASCII text format or as ASCII delimited database files.

Data can be accepted on 5.25 or 3.5 inch double-density or high-density floppy disks or CD-ROM, over e-mail, or via FTP. If data are transferred via FTP we prefer to 'pull' data, so please contact the ADS in advance to make arrangements. Other file formats can be accommodated, please e-mail the ADS for details.

6.3.4 Encoding files

We prefer data not to be encoded, but can accept UUENCODED files.

6.3.5 Compressing files

If data is compressed, we prefer the following formats: GNU (.gz), Pkzip, Stuffit, TAR, Unix compressed files (.Z), or Zip.

In some cases, data sets may need to be compressed to assist in transferring information to the archive or out of it again. Large data sets with numerous component files are most easily handled in this way. For example, if you are using GRASS the complete directory structure associated with a location or mapset spatial database might be transferred to the ADS as a single compressed file.

6.3.6 If you have a problem or question...

More detailed information about depositing information with the Archaeology Data Service is located in the Guidelines for Depositors. Please feel free to contact the ADS directly if you have any questions or concerns.

Archaeology Data Service
University of York
King's Manor
York YO1 7EP
Telephone: (01904) 433 954
Fax: (01904) 433 939
Email: info@ads.ahds.ac.uk
http://ads.ahds.ac.uk/

6.4 CASE STUDY – DEPOSITING A GIS DATA SET

6.4.1 Introduction to the Case Study

As discussed in the previous section data deposited for archiving should be accompanied by appropriate documentation and a signed license form. To illustrate this process, a simple case study based on data from a real archaeological project is provided.

Data prepared for deposit consisted mainly of Arc/Info export files and some postscript images showing key coverages in the GIS. The importance of documentation becomes immediately clear – imagine if you were given this data, and the only thing you knew about it was that it consisted mainly of Arc/Info export files!

6.4.2 Documentation

Luckily the depositor, Dr Julian Richards of the University of York, provided a WordPerfect text file in addition to his export files:

BACKGROUND INFORMATION ABOUT COTTAM PROJECT

BY JULIAN RICHARDS

The presence of a settlement site of the 8th – 10th centuries AD in the parish of Cottam and Cowlam, East Yorkshire, was first indicated by the distribution of over 200 metal objects discovered from 1987 onwards by metal detector users working fields to the west of Burrow House Farm. The majority of the metal finds were published in the *Yorkshire Archaeological Journal* (Haldenby 1990, 1992, 1994), although at that stage the location of the site was not revealed because of the dangers of unauthorised metal detecting. The original group of metal detector users had recorded the approximate location of each of their finds and their distribution was shown to correspond with a sub-rectangular crop mark enclosure. Rural sites which can be assigned to the Anglian and Anglo-Scandinavian period are rare in England and it was decided to carry out an archaeological evaluation of the site, using field walking, geophysics and trial excavation (Richards 1994). Three seasons of excavation were carried out, between 1993–96, sponsored by the Department of Archaeology, University of York, Earthwatch, and the British Academy. The aims of the evaluation were to:

1.   examine the relationship between the crop marks and the 8th–10th century finds
2.   characterise the nature of the settlement activity
3.   assess the extent of plough damage and the survival of occupation layers
4.   recover artefactual and environmental samples

The role of the GIS was in the integration of the different categories of evidence and to help understand the spatial development of the site both internally and within its wider landscape setting (Richards 1996). It is believed that this question is critical for our understanding of the development of early medieval settlement patterns in Northumbria. Within ARC/INFO coverages were defined for several classes of data, including:

1.   the distribution of metal artefacts recorded by the metal-detectorists;
2.   the aerial photographic coverage;
3.   data collected in two seasons of field walking;
4.   magnetometer and resistivity survey data;
5.   evidence from the trial excavation.

These data incorporate point, line and polygon information. Associated attribute tables include detailed finds information which enable the development of artefact distribution patterns, incorporating those finds recovered by the metal detector users, and during the controlled excavations. The final excavation report is in preparation (Richards in prep).

Haldenby, D. 1990 'An Anglian site on the Yorkshire Wolds', *Yorkshire Archaeological Journal* **62**, 51–63
____   1992 'An Anglian site on the Yorkshire Wolds', *Yorkshire Archaeological Journal* **64**, 25–39
____   1994 'An Anglian site on the Yorkshire Wolds – Part III', *Yorkshire Archaeological Journal* **66**,
Richards, J.D. 1994 'Cottam Evaluation' *Yorkshire Archaeological Journal* **66**, 57–8
____   1996 'Putting the site in its setting: GIS and the search for Anglo-Saxon settlements in Northumbria'
          Analecta Praehistorica Leidensia **28**, 377–86
Richards, J.D. in prep 'Cottam: An Anglian and Anglo-Scandinavian Settlement on the Yorkshire Wolds'

6.4.3 Data and Metadata

This background documentation about the project accompanied the actual data sets. Only the image files will be discussed in the remainder of this case study.

An original image file showing the main aerial photograph interpretation for Cottam was deposited as a postscript file:

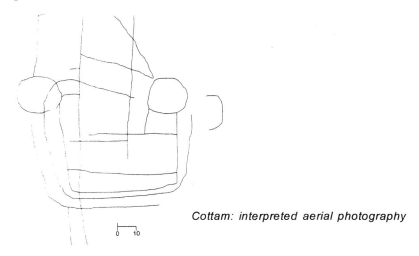

Cottam: interpreted aerial photography

Metadata supplied at the time of deposit gives useful background information about this image:

Dublin Core Metadata for Image	
Title	Cottam B Enclosure aerial photograph transcription
Creator	John Duffy interpreted and transcribed image
Subject	settlement, excavation, survey, aerial photography, Anglo-Saxon
Description	This aerial photograph interpretation was created during the post-excavation analysis of the Cottam B Project directed by Julian Richards. Image was imported into Arc/Info from Autocad. This Anglo-Saxon settlement is located in Yorkshire on the Wolds east of York.
Publisher	The image is licensed for distribution by the Archaeology Data Service
Date	1998
Type	image
Format	postscript (eps) and gif
Identifier	http://ads.ahds.ac.uk/images/ap.gif
Source	Aerial photograph held by the Royal Commission on the Historical Monuments of England
Language	English
Relation	Derived from a larger GIS data set held by Julian Richards, Department of Archaeology, University of York. Data is from site 6865 in the Humberside Sites and Monuments Record. Site is called "Burrow House Farm, Cottam" in the RCHME National Excavation Index for England.
Coverage – Spatial	4976 4667
Coverage – Temporal	Early Medieval – this term is derived from a controlled MIDAS list (RCHME 1998)
Rights	Image is freely accessible to registered Archaeology Data Service users.

There was also an image file showing the trench locations for the Cottam excavation:

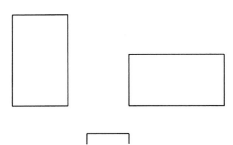

Metadata supplied at the time of deposit gives useful background information about this image:

Dublin Core Metadata for Image	
Title	Cottam 1993 excavation trench outlines
Creator	Julian Richards
Subject	settlement, excavation, Anglo-Saxon
Description	The location of excavation trenches for the 1993 Cottam excavation are shown in this image. Data was collected using an EDM, and then downloaded into Arc/Info. This Anglo-Saxon settlement is located in Yorkshire on the Wolds east of York.
Publisher	The image is licensed for distribution by the Archaeology Data Service
Date	1995
Type	image
Format	postscript (eps) and gif
Identifier	http://ads.ahds.ac.uk/images/trench2.gif
Source	EDM survey
Language	English
Relation	Derived from a larger GIS data set held by Julian Richards, Department of Archaeology, University of York. Data is from site 6865 in the Humberside Sites and Monuments Record. Site is called "Burrow House Farm, Cottam" in the RCHME National Excavation Index for England.
Coverage – Spatial	4976 4667
Coverage – Temporal	Early Medieval – this term is derived from a controlled MIDAS list (RCHME 1998)
Rights	Image is freely accessible to registered Archaeology Data Service users.

The third and final original image file showed the location of metal artefacts found at Cottam:

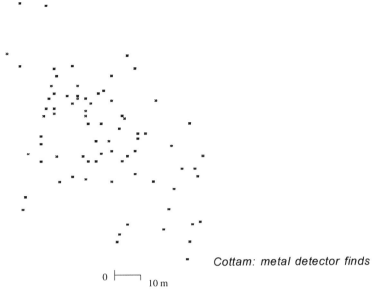

Cottam: metal detector finds

0 ├───┤ 10 m

Metadata supplied at the time of deposit again gives useful background information about this image:

Dublin Core Metadata for Image	
Title	Cottam B metal detector finds
Creator	Dave Haldenby conducted the survey, Tony Austin entered data, and Julian Richards managed the GIS
Subject	settlement, excavation, survey, metal objects, coins, Anglo-Saxon
Description	The metal objects found by metal detector were plotted as part of the post-excavation analysis of the Cottam B Project, directed by Julian Richards. Data was entered into a Paradox database and then exported to Arc/Info. Cottam is an Anglo-Saxon settlement located in Yorkshire on the Wolds east of York.
Publisher	The image is licensed for distribution by the Archaeology Data Service
Date	1995
Type	image
Format	postscript (eps) and gif
Identifier	http://ads.ahds.ac.uk/images/metal-de.gif
Source	Metal detector survey
Language	English
Relation	Derived from a larger GIS data set held by Julian Richards, Department of Archaeology, University of York. Data is from site 6865 in the Humberside Sites and Monuments Record. Site is called "Burrow House Farm, Cottam" in the RCHME National Excavation Index for England.
Coverage – Spatial	4976 4667
Coverage – Temporal	Early Medieval – this term is derived from a controlled MIDAS list (RCHME 1998)
Rights	Image is freely accessible to registered Archaeology Data Service users.

As mentioned in section 6.3, tools are being developed by the ADS to assist in the creation of metadata. This will facilitate the creation of standard, structured metadata rather than free-text entries for each of the fields – important when using metadata to enable searching for data sets.

6.4.4 Too much work?

If you were considering deposit of an entire GIS data set you might not want to create such detailed metadata records for every single layer. There is a high degree of overlap in the metadata for these three images, and this points to another possible option – creating a single metadata record describing the contents of the **entire** GIS data set that you wished to archive. In this case it would be important to use enough subject keyword terms to identify themes within the GIS data set.

Section 7: Useful definitions and references

7.1 SELECTED BIBLIOGRAPHY

Aldenderfer, M. and Maschner, H.D.G. (eds) 1996. *Anthropology, Space, and Geographic Information Systems*. New York: Oxford University Press.

Allen, K.M.S., Green, S.W. and Zubrow, E.B.W. (eds) 1990. *Interpreting Space: GIS and archaeology*. London: Taylor & Francis.

Andresen, J., Madsen, T. and Scollar, I. (eds) 1993. *Computing the Past. Computer Applications and Quantitative Methods in Archaeology, CAA92*. Aarhus: Aarhus University Press.

Aronoff, S. 1989. *Geographical Information Systems: A Management Perspective*. Ottawa: WDL Publications.

Arroyo-Bishop, D. and Lantada Zarzosa, M.T. 1995. To be or not to be: will an object-space-time GIS/AIS become a scientific reality or end up an archaeological entity? In G. Lock and Z. Stancic (eds) *Archaeology and Geographical Information Systems: a European Perspective*: 43–54. London: Taylor & Francis.

Batini, C., Ceri, S. and Navathe, S. 1992. *Conceptual database design: an entity-relationship approach*. Benjamin Cummins.

Bewley, R., Donoghue, D., Gaffney, V., van Leusen, M. and Wise, A. 1999 Archiving Aerial Photography and Remote Sensing Data: A Guide to Good Practice. Oxbow. Oxford.

Biswell, S., Cropper, L, Evans, J., Gaffney, V. and Leach, P. 1995. GIS and excavation: a cautionary tale from Shepton Mallett, Somerset, England. In G. Lock and Z. Stancic (eds) *Archaeology and Geographical Information Systems: a European Perspective*: 269–285. London: Taylor & Francis.

Brandt, R., Groenewoudt, B.J. and Kvamme, K.L. 1992. An experiment in archaeological site location: modelling in the Netherlands using GIS techniques. *World Archaeology* 24: 268–282.

Bugayevsky, L.M. and Snyder, J.P. 1995. *Map Projections: A Reference Manual*. London: Taylor & Francis.

Burrough, P.A. 1986. *Principles of Geographical Information Systems for Land Resources Assessment*. Oxford: Clarendon Press.

Castleford, J. 1992. Archaeology, GIS and the time dimension: an overview. In G. Lock, and J. Moffett (eds) *Computer Applications and Quantitative Methods in Archaeology 1991*: 95–106. Oxford: Tempus Reparatum, British Archaeological Reports International Series S577.

Clancy, J. 1991. *Site surveying and levelling*. London: Edward Arnold.

Claxton, J.B. 1995. Future enhancements to GIS: implications for archaeological theory. In G. Lock and Z. Stancic (eds) *Archaeology and Geographical Information Systems: a European Perspective*: 335–348. London: Taylor & Francis.

Cox, C. 1992. Satellite imagery, aerial photography and wetland archaeology. *World Archaeology* 24 (2): 249–67.

Date, C.J. 1995. *An introduction to database systems, 6th edition*. Addison-Wesley.

Defence Mapping Agency, 1984. *Geodesy for the Layman*. US Naval Observatory Report TR 80–003, Washington D.C., 16th March 1984.

DeMers, M.N. 1997. *Fundamentals of Geographic Information Systems*. New York, John Wiley & Sons.

ESRI 1996. *The GIS Glossary*. Environmental Systems Research Institute, Inc

Evenden G.I. 1983. *Forward and inverse cartographic projection procedures*. US Geological Survey Open-File Report 83–623.

Evenden G.I. 1990. *Cartographic projection procedures for the UNIX environment – a user's manual*. US Geological Survey Open-File Report 90–284.

Fotheringham, S. and Rogerson, P. (eds) 1994. *Spatial analysis and GIS*. London: Taylor & Francis.

Gaffney, V. and Stancic, Z. 1991. *GIS approaches to regional analysis: A case study of the island of Hvar*.

Ljubljana: Znanstveni inštitut Filozofske fakultete, University of Ljubljana, Yugoslavia, (reprinted 1996).

Gaffney, V. and Stancic, Z. 1992. Diodorus Siculus and the island of Hvar, Dalmatia: testing the text with GIS. In G. Lock, and J. Moffett (eds) *Computer Applications and Quantitative Methods in Archaeology 1991*: 113–125. Oxford: Tempus Reparatum, British Archaeological Reports International Series S577.

Gaffney, V. and van Leusen, P.M. 1995. Postscript – GIS, environmental determinism and archaeology. In G. Lock and Z. Stancic (eds) *Archaeology and Geographical Information Systems: a European Perspective*: 367–382. London: Taylor & Francis.

Gaffney, V., Stancic, Z. and Watson, H. 1995. The impact of GIS on archaeology: a personal perspective. In G. Lock and Z. Stancic (eds) *Archaeology and Geographical Information Systems: a European Perspective*: 211–230. London: Taylor & Francis.

Gaffney, V., van Leusen, M., and White, R. 1996. Mapping the Past: Wroxeter and Romanisation, in T. Higgins, P. Main, and J. Lang, (eds) 1996. *Imaging The Past. Electronic Imaging and Computer Graphics in Museums and Archaeology*: 259–270. London: British Museum Occasional Paper Number 114.

Gaffney, V., Ostir, K., Podobnikar, T. and Stancic, Z. 1996. Satellite imagery and GIS applications in Mediterranean landscapes. In H. Kamermans and K. Fennema (eds) *Interfacing the Past. Computer Applications and Quantitative Methods in Archaeology CAA95*: 337–342. Leiden: University of Leiden, Analecta Praehistorica Leidensia 28. (2 volumes).

Gillings, M. 1995. Flood dynamics and settlement in the Tisza valley of north-east Hungary: GIS and the Upper Tisza project. In G. Lock and Z. Stancic (eds) *Archaeology and Geographical Information Systems: a European Perspective*: 67–84. London: Taylor & Francis.

Gillings, M. and Goodrick, G.T. 1996. Sensuous and reflexive GIS: exploring visualisation and VRML. *Internet Archaeology* 1 [online]. Available from http://intarch.ac.uk/journal/issue1/gillings_index.html.

Guillot, D. and Leroy, G. 1995. The use of GIS for archaeological resource management in France: the SCALA Project. In G. Lock and Z. Stancic (eds) *Archaeology and Geographical Information Systems: a European Perspective*: 15–26. London: Taylor & Francis.

Gurtner, W. 1994. *RINEX: The receiver Independent Exchange Format Version*. Available from http://www.eng.auburn.edu/~yoonseo/gpswww/documents/rinex2.txt

Gurtner, W. and Mader, G.M. 1990. The RINEX Format: *Current Status, Future Developments*. Available from http://www.eng.auburn.edu/~yoonseo.gpswww/documents/rinex.txt

Harris, T.M. and Lock, G.R. 1990. The diffusion of a new technology: a perspective on the adoption of geographic information systems within UK archaeology, in K.M.S. Allen, S.W. Green and E.B.W. Zubrow (eds) *Interpreting Space: GIS and archaeology*: 33–53. London: Taylor & Francis.

Harris, T. and Lock, G. 1995. Toward an evaluation of GIS in European archaeology: the past, present and future of theory and applications. In G. Lock and Z. Stancic (eds) *Archaeology and Geographical Information Systems: a European Perspective*: 349–365. London: Taylor & Francis.

Harris, T.M. and Lock, G.R. 1996. Multi-dimensional GIS: exploratory approaches to spatial and temporal relationships within archaeological stratigraphy. In H. Kamermans and K. Fennema (eds) *Interfacing the Past. Computer Applications and Quantitative Methods in Archaeology CAA95*: 307–316. Leiden: University of Leiden, Analecta Praehistorica Leidensia 28. (2 volumes).

HMSO 1988. *The Copyright, Designs and Patents Act*. London: Her Majesty's Stationery Office.

Huggett, J. and Ryan, N. (eds) 1995. *Computer Applications and Quantitative Methods in Archaeology 1994*. Oxford: Tempus Reparatum, British Archaeological Reports International Series 600.

Kamermans, H. and Fennema, K. (eds) 1996. *Interfacing the Past. Computer Applications and Quantitative Methods in Archaeology CAA95*. Leiden: University of Leiden, Analecta Praehistorica Leidensia 28. (2 volumes).

Kohler, T.A. and Parker, S.C. 1986. Predictive models for archaeological resource location. In M. B. Schiffer (ed.) *Advances in Archaeological Method and Theory, Vol. 9*: 397–452. New York: Academic Press.

Kvamme, K.L. 1990. The fundamental principles and practice of predictive modelling. In A. Voorrips (ed) 1990. *Mathematics and Information Science in Archaeology: a Flexible Framework*: 257–295. Bonn: Studies in Modern Archaeology 3, Holos-Verlag.

Kvamme, K.L. 1993. Spatial statistics and GIS: an integrated approach, in J. Andresen, T. Madsen and I. Scollar (eds) *Computing the Past. Computer Applications and Quantitative Methods in Archaeology, CAA92*: 91–103. Aarhus: Aarhus University Press.

Kvamme, K.L. 1995. A view from across the water: the North American experience in archaeological GIS. In G.

Lock and Z. Stancic (eds) *Archaeology and Geographical Information Systems: a European Perspective*: 1–14. London: Taylor & Francis.

Kvamme, K.L. 1996. Investigating chipping debris scatters: GIS as an analytical engine, in H.D.G. Maschner (ed) *New Methods, Old Problems. Geographic Information Systems in modern archaeological research*: 38–71. Carbondale: Southern Illinois University Center for Archaeological Investigations. Occasional Paper No. 23.

Kvamme, K.L. and Kohler, T.W. 1988. Geographic information systems: technical aids for data collection, analysis and display, in J.W. Judge and L. Sebastian (eds) *Quantifying the present and Predicting the Past: Theory, Method and Application of Archaeological Predictive Modelling*: 493–548. Washington DC: US Bureau of Land Management, Department of Interior, US Government Printing Office

Laflin, S. (ed.) 1986. *Computer Applications in Archaeology 1986*. Birmingham: Centre for Computing and Computer Science, University of Birmingham.

Lang, N. and Stead, S. 1992. Sites and Monuments Records in England – theory and Practice, in G. Lock, and J. Moffett (eds) *Computer Applications and Quantitative Methods in Archaeology 1991*: 69–76. Oxford: Tempus Reparatum, British Archaeological Reports International Series S577.

Larsen, C.U. (ed.), 1992. *Sites and Monuments. National Archaeological Records*. Copenhagen: The National Museum of Denmark, DKC.

Laurini, R. and Thompson, D. 1996. *Fundamentals of Spatial Information Systems*. London: Academic Press.

Leick, A. 1995. *GPS Satellite Surveying, 2nd Edition*. New York: John Wiley & Sons.

Lillesand ,T.M. and Keifer, R.W. 1994. *Remote sensing and image interpretation, third edition*. New York: Wiley & Sons.

Llobera, M. 1996. Exploring the topography of mind: GIS, social space and archaeology. *Antiquity* 70: 612–22.

Lock, G.R. and Harris, T.M. 1996. Danebury revisited: an English Iron Age hillfort in a digital landscape, in M. Aldenderfer and H.D.G. Maschner (eds) *Anthropology, Space, and Geographic Information Systems*: 214–240. New York: Oxford University Press.

Lock, G. and Moffett, J. (eds) 1992. *Computer Applications and Quantitative Methods in Archaeology 1991*. Oxford: Tempus Reparatum, British Archaeological Reports International Series S577.

Lock, G. and Stancic, Z. (eds) 1995. *Archaeology and Geographical Information Systems: a European Perspective*. London: Taylor & Francis.

Lockyear, K. and Rahtz, S. (eds) 1991. *Computer Applications and Quantitative Methods in Archaeology 1990*. Oxford: Tempus Reparatum, British Archaeological Reports International Series 565.

Lyall, J. and Powlesland, D. 1996. The application of high resolution fluxgate gradiometry as an aid to excavation planning and strategy formulation. *Internet Archaeology* 1 [online]. Available from http://intarch.ac.uk/journal/issue1/lyall_index.html.

Martin, D. 1996. *Geographic Information Systems. Socioeconomic applications (second edition)*. London: Routledge.

Maschner, H.D.G. 1996. Geographic Information Systems in Archaeology, in H.D.G. Maschner (ed) *New Methods, Old Problems. Geographic Information Systems in modern archaeological research*: 1–21. Carbondale: Southern Illinois University Center for Archaeological Investigations. Occasional Paper No. 23.

Maschner, H.D.G. (ed) 1996. *New Methods, Old Problems. Geographic Information Systems in modern archaeological research*. Carbondale: Southern Illinois University Center for Archaeological Investigations. Occasional Paper No. 23.

mda 1997. *MDA Archaeological Objects Thesaurus*. Cambridge: Museum Documentation Association, English Heritage, and the Royal Commission for the Historical Monuments of England.

Miller, P. and Greenstein, D. 1997. *Discovering Online Resources Across the Humanities: A Practical Implementation of the Dublin Core*. Bath: UKOLN.

Mobbs, R. 1997. *Glossary of Computing Terms* [online]. Available from http://www.le.ac.uk/cc/iss/glossary/ccgl.html.

Murray, D.M. 1995. The management of archaeological information – a strategy, in J. Wilcock and K. Lockyear (eds) *Computer Applications and Quantitative Methods in Archaeology 1993*: 83–87. Oxford: Tempus Reparatum, British Archaeological Reports International Series S598.

Murray, D. and Dixon, P. 1995. Geographic Information System for RCAHMS. In *Monuments on Record, Annual Review 1994–5 of the Royal Commission on the Ancient and Historical Monuments of Scotland*: 32–4.

NMEA 1995. *NMEA 0183 Standard*. National Maritime Electronics Association, PO Box 3435, New Bern, NC 28564–3435, USA.

Openshaw, S. 1991. Developing appropriate spatial analysis methods for GIS. In D.J. Maguire, M.F. Goodchild and D.W. Rhind (eds) *Geographical Information Systems, Vol. 2*: 389–402.

Ordnance Survey. 1950. *Constants, formulae and methods used in the Transverse Mercator Projection*. London: HMSO.

Ordnance Survey. 1995. *The ellipsoid and the Transverse Mercator Projection*. Geodetic Information Paper No 1. Southampton: Ordnance Survey.

Ordnance Survey. 1996. *Geodetic Information Paper No 2*. Southampton: Ordnance Survey.

Ordnance Survey. 1996. *Copyright leaflet #1*. Southampton: Ordnance Survey.

Ordnance Survey. 1996. *Copyright leaflet #2*. Southampton: Ordnance Survey.

Ordnance Survey. 1996. *Copyright leaflet #3*. Southampton: Ordnance Survey.

Ordnance Survey. 1996. *Copyright leaflet #4*. Southampton: Ordnance Survey.

Rahtz, S.P.Q. 1988. *Computer Applications and Quantitative Methods in Archaeology 1988*. Oxford: British Archaeological Reports International Series 446 (2 volumes).

Rahtz, S.P.Q. and Richards, J. (eds), 1989. *Computer Applications and Quantitative Methods in Archaeology 1989*. Oxford: British Archaeological Reports International Series 548.

Raper, J.F. (ed.) 1989. *Three dimensional applications in geographical information systems*. London: Taylor & Francis.

RCHME, 1998. *Thesaurus of Monument Types*. Swindon: Royal Commission on the Historical Monuments of England. Second edition.

RCHME, 1998. *MIDAS: the Monument Inventory Data Standard*. Swindon: Royal Commission on the Historical Monuments of England.

Richards, J.D. and Ryan, N.S. 1985. *Data Processing in Archaeology*. Cambridge: Cambridge University Press.

Robinson, H. 1993. The archaeological implications of a computerised integrated National Heritage Information System. In J. Andresen, T. Madsen, and I. Scollar (eds) *Computing the Past. Computer Applications and Quantitative Methods in Archaeology, CAA92*: 139–150. Aarhus: Aarhus University Press.

Roorda, I.M. and Wiemer, R. 1992. Towards a new archaeological information system in the Netherlands. In G. Lock, and J. Moffett (eds) *Computer Applications and Quantitative Methods in Archaeology 1991*: 85–88. Oxford: Tempus Reparatum, British Archaeological Reports International Series S577.

Ruggles, C. 1992. Abstract Data Structures for GIS Applications in Archaeology. In G. Lock, and J. Moffett (eds) *Computer Applications and Quantitative Methods in Archaeology 1991*: 107–112. Oxford: Tempus Reparatum, British Archaeological Reports International Series S577.

Ruggles, C.L.N. and Rahtz, S.P.Q. (eds) 1988. *Computer Applications and Quantitative Methods in Archaeology 1987*. Oxford: British Archaeological Reports, International Series 393.

Ryan, N.S. and Smith, D.J. 1995. *Database systems engineering*. London: International Thompson Computer Press.

Scollar, I., Tabbagh, A. and Hesse, A. 1990. *Archaeological prospecting and remote sensing*. Cambridge: Cambridge University Press.

Snyder, J.P. 1987. *Map projections – a working manual*. U.S. Geological Survey Professional Paper 1395.

Snyder, J.P. 1989. *An Album of Map Projections*. U.S.Geological Survey Professional Paper 1453.

Stead, S. 1995. Humans and PETS in space. In G. Lock and Z. Stancic (eds) *Archaeology and Geographical Information Systems: a European Perspective*: 313–318. London: Taylor & Francis.

Taylor, P.J. and Johnston, R.J. 1995. GIS and Geography. In J. Pickles (ed) *Ground Truth. The social implications of Geographic Information Systems*: 51–67. London: The Guilford Press.

Tilley, C. 1994. *A phenomenology of landscape. Places, Paths and Monuments*. Oxford: Berg.

Tomlin, C.D. 1990. *Geographic Information Systems and Cartographic Modelling*. New Jersey: Prentice-Hall.

van Leusen, M. 1993. Cartographic modelling in a cell-based GIS. In J. Andresen, T. Madsen, and I. Scollar (eds) *Computing the Past. Computer Applications and Quantitative Methods in Archaeology, CAA92*: 105–124. Aarhus: Aarhus University Press.

van Leusen, M. 1995. GIS and archaeological resource management: a European agenda. In G. Lock and Z. Stancic (eds) *Archaeology and Geographical Information Systems: a European Perspective*: 27–42. London: Taylor & Francis.

van Leusen, P.M. 1996. GIS and locational modeling in Dutch archaeology: a review of current approaches. In H.D.G. Maschner (ed.) *New Methods, Old Problems. Geographic Information Systems in modern archaeological research*: 177–197. Carbondale: Southern Illinois University Center for Archaeological Investigations. Occasional Paper No. 23.

van Sickle, J. 1996. *GPS for Land Surveyors*. Ann Arbor Press.

Verhagen, P. 1996. The use of GIS as a tool for modelling ecological change and human occupation in the middle Aguas Valley (S.E. Spain). In H. Kamermans, and K. Fennema, (eds) *Interfacing the Past. Computer Applications and Quantitative Methods in Archaeology CAA95*: 317–324. Leiden: University of Leiden, Analecta Praehistorica Leidensia 28. (2 volumes).

Walker, R. (ed.) 1993. *AGI Standards Committee GIS Dictionary*. Association for Geographic Information.

Warren, R.E. 1990. Predictive modelling of archaeological site location: a primer. In K.M.S. Allen, S.W. Green, and E.B.W. Zubrow, (eds) *Interpreting Space: GIS and archaeology*: 90–111. London: Taylor & Francis.

Wells, D. 1986. *Guide to GPS Positioning*. Canadian GPS Associates.

Wheatley, D. W. 1993. Going over old ground: GIS, archaeological theory and the act of perception. In J. Andresen, T. Madsen, and I. Scollar (eds) *Computing the Past. Computer Applications and Quantitative Methods in Archaeology, CAA92*: 133–138. Aarhus: Aarhus University Press.

Wheatley, D.W. 1995. Cumulative Viewshed Analysis: a GIS-based method for investigating intervisibility, and its archaeological application. In G. Lock and Z. Stancic (eds) *Archaeology and Geographical Information Systems: a European Perspective*: 171–186. London: Taylor & Francis.

Wheatley, D.W. 1996. Between the lines: the role of GIS-based predictive modelling in the interpretation of extensive survey data. In H. Kamermans, and K. Fennema (eds) *Interfacing the Past. Computer Applications and Quantitative Methods in Archaeology CAA95*: 275–292. Leiden: University of Leiden, Analecta Praehistorica Leidensia 28. (2 volumes).

Whittington, R.P. 1988. *Database Systems Engineering*. Oxford: Oxford Applied Mathematics and Computing Science Series.

Wilcock, J. and Lockyear, K. (eds) 1995. *Computer Applications and Quantitative Methods in Archaeology 1993*. Oxford: Tempus Reparatum, British Archaeological Reports International Series S598.

Wise, A.L. and Miller, P. 1997. Why Metadata Matters in Archaeology. In *Internet Archaeology* 2 [online]. Available from http://intarch.ac.uk/journal/issue2/wise_index.html

7.2 GLOSSARY

2.5D   Most GIS do not actually achieve full 3-dimensional representations of landscapes, but instead render representations in 2.5 dimensions. This means that an attribute value is used instead of a full spatial location in the z axis. DTMs are an example of 2.5 dimensional representations.

ADS   The Archaeology Data Service.

AGI   The Association for Geographic Information.

Arc/Info   A commercial GIS package widely used within academia. The UNIX and NT releases provide comprehensive raster and vector processing capabilities. The PC release is vector only.

ArcView   A commercial desktop mapping/GIS package widely used within academia.

BIL   Band Interleaved by Line. An image file format linked with satellite derived imagery.

BMP   BitMaP. A file extension indicating a graphics file, common in Windows applications.

CAA   Computer Applications in Archaeology. An annual international conference that has been instrumental in developing GIS applications within archaeology.

CAD   Computer Aided Design. The design activities, including drafting and illustrating, in which information processing systems are used to carry out functions such as designing or improving a part or a product (Walker 1993).

CBA   The Council for British Archaeology.

CIDOC   The International Documentation Committee of the International Council of Museums.

CGM   Computer Graphics Metafile. A standard (ISO 8632) file format specification for the storage and transfer of picture description information (Walker 1993).

DBASE   A commercial relational database system. Widely used within archaeology.

DBF   DataBase File. A proprietary database file format used by DBASE. Often used as a de facto standard to exchange database files.

DEM   Digital Elevation Model. The term DEM can refer to one of the following:
1. A digital representation of a continuous variable over a two- dimensional surface by a regular array of z values referenced to a common datum. Digital elevation models are typically used to represent terrain relief and frequently comprise a foundational layer in any archaeological GIS database.
2. An elevation database for elevation data by map sheet from the National Mapping Division of the U.S. Geological Survey (USGS).
3. The format of the USGS digital elevation data sets (ESRI 1996)

DLG   Digital Line Graph. The digital format standards published by US Geological Survey for exchanging cartographic data files and in which the USGS delivers topographical maps in vector format (Walker 1993).
1. Digital Line Graph files from the U.S. Geological Survey (USGS), including data from the base map categories such as transportation, hydrography, contours, and public land survey boundaries.
2. The digital format standards published by USGS for exchanging cartographic data files and in which the USGS delivers Digital Line Graph data sets (ESRI 1996).

DSM   Digital Surface Model. Largely synonymous with a DEM but with the added possibility of being a component in a stack of surface models.

DTM   Digital Terrain Model. A term which is commonly used interchangeably with DEM. Strictly speaking a DTM refers to a model of reality which includes information relating to factors such as surface texture as well as elevation.

Dublin Core A 15 field standard for metadata – or "information about information". More information is available on the Web at http://purl.oclc.org/metadata/dublin_core.

DXF   Drawing eXchange Format. A format for transferring drawings between Computer Aided Design systems, widely used as a de facto standard in the engineering and construction industries (Walker 1993).

ECEF   Earth-Centred, Earth-Fixed. A Cartesian co-ordinate system used by satellite positioning systems, aligned with the WGS 84 reference ellipsoid.

EDM   Electronic Distance Measure. Digital measuring device used within terrestrial survey. It is based upon the transit time measurement of an electromagnetic beam emitted from a transmitter/receiver to a reflecting target prism and back again (Clancy 1991: 285). Often incorrectly used by archaeologists to identify Total Station Integrated survey instruments, one component of which is an integral EDM.

ESRC   Environmental and Social Research Council in the United Kingdom.

FGDC   The United States Federal Geographic Data Committee. Composed of representatives of several federal agencies and GIS vendors, the FGDC has the lead role in defining spatial metadata standards, which it describes in the Content Standards for Spatial Metadata (ESRI 1996).

FTP   File Transfer Protocol.

GCP   Ground Control Point. A point on the surface of the earth of known location (i.e. fixed within an established co-ordinate system) which is used to geo-reference Image data sources, such as remotely sensed images or scanned maps, and divorced survey grids, such as those generated during geophysical survey.

GDOP   Geometric Dilution Of Precision. Used within satellite-based survey as a measure of the quality of the fix indicating the suitability of satellite positions for triangulation.

Geotiff An extension to the TIF graphics standard to incorporate georeferencing information. Although currently supported by a limited number of proprietary GIS, many manufacturers have committed to supporting the standard. It aims to provide a platform-independent method for archiving and tranbsferring spatially referenced raster products.

GIF   Graphics Interchange Format. A bitmap graphics format from CompuServe which stores screen images economically and aims to maintain their correct colours even when transferred between different computers (Mobbs 1997).

GIS   Geographic Information System. An organised collection of computer hardware, software, geographic data, and personnel designed to efficiently capture, store, update, manipulate, analyse, and display all forms of geographically referenced information (ESRI 1996).

A computer system for capturing, storing, checking, integrating, manipulating, analysing and displaying data related to positions on the Earth's surface. Typically, a Geographical Information System (or Spatial Information System) is used for handling maps of one kind or another. These might be represented as several different layers where each layer holds data about a particular kind of feature. Each feature is linked to a position on the graphical image of a map (Walker 1993)

GLONASS GLObal'naya NAvigatsionnaya Sputnikovaya Sistema. The Russian Global Navigation Satellite System is currently a constellation of 53 spacecraft deployed in nearly semi-synchronous orbits. The Phase I constellation was established in 1990; a 21-spacecraft constellation is the operational goal.

GNU   Gnu's Not Unix. A project providing free versions of unix and a large number of freeware tools, one of which is GNU Zip (.gz format files).

GPS   Global Positioning System. A satellite based navigational system allowing the determination of any point on the earth's surface with a high degree of accuracy given a suitable GPS receiver (Walker 1993).

GRASS   Geographic Resources Analysis Support System. This is a public-domain raster GIS modelling product of the US Army Corp. of Engineers Construction Engineering Research Laboratory (Walker 1993). It is in common use within archaeology.

GRID   The raster module of the Arc/Info GIS package.

HRV   High Resolution Visible. This is a specific sensor carried aboard the SPOT satellite capable of achieving a spatial resolution of 10 metres (Walker 1993).

HTML   HyperText Markup Language. The general framework for defining document structure used with the World Wide Web facility of the Internet.

IDRISI   A raster-based commercial GIS package in common use amongst archaeologists.

JPEG   Joint Photographic Expert Group. The original name of the committee that designed the standard image compression algorithm. JPEG is designed for compressing either full colour or grey-scale digital images of 'natural', real-world scenes. It does not work so well on non-realistic images, such as cartoons or line drawings. JPEG does not handle compression of black-and-white (1-bit-per-pixel) images or moving pictures (Walker 1993).

Landsat   A series of satellites that produce images of the earth. The Landsat remote sensing satellite program was developed by NASA (National Aeronautics and Space Administration). Landsat data are provided in .BIL (band interleaved by line) or .BIP (band interleaved by pixel) formats (ESRI 1996).

LaTeX   A widely used document exchange format.

mda   Museum Documentation Association.

MIDAS   Monument Inventory Data Standard.

MIF/MID   Mapinfo export formats.

MOSS   This is a public domain GIS developed by the U.S. Department of Interior.

NERC   The Natural Environment Research Council.

NGDF   National Geospatial Data Framework. An Important co-operative initiative which aims to provide effective means of accessing geospatial data collected and held by government and the public/private sectors.

NMEA   National Maritime Electronics Association. An organisation involved in the development of output protocols for satellite receivers.

NMR   National Monument Record.

NTF   National Transfer Format. An implementation of British Standard BS7567, used for the transfer of geographic data. It is administered by the Association for Geographic Information in the United Kingdom.

ODA   A documentation exchange format.

OS   The Ordnance Survey. Great Britain's national mapping agency.

OSGB   The Ordnance Survey of Great Britain.

PDF   Portable Document Format. A document standard promoted by Adobe.

PhotoCD   An image exchange format promoted by Kodak.

PKZip   A file compression utility usually found on PC systems.

PNG   Portable Network Graphics. Pronounced 'ping' The PNG format is intended to provide a portable, legally unencumbered, well-compressed, well-specified standard for lossless bitmapped image files. Although the initial motivation for developing PNG was to replace GIF, the design provides some useful new features not available in GIF, with minimal cost to developers (http://www.eps.mcgill.ca/~steeve/PNG/png.html).

RCHME   The Royal Commission on the Historical Monuments of England.

RINEX   Receiver INdependent EXchange Format. A widely used satellite receiver output protocol, not tied into any particular device or class of device.

RMS   Root Mean Square. This is an error measurement that most GIS report during geometric transformation of data sets. It is mathematically the spatial equivalent to the standard deviation. The RMS error is often used as a measure of the accuracy of tic points when registering a map to a digitiser, indicating the discrepancy between known point locations and their digitised locations. The lower the RMS error, the more accurate the digitising or transformation (Walker 1993). The fact that the RMS error is expressed as one simple figure (e.g. 5.67 m) unfortunately does not mean that any point in the transformed image will be within this distance from its 'real' coordinates. In fact, the actual error can vary across the image depending on the number, placement, and accuracy of the tiepoints used.

RTF   Rich Text Format. A widely used document exchange format.

SAR   Synthetic Aperture Radar. A satellite based technique for generating a regular matrix of elevation values (i.e. a DEM) directly.

SDTS   Spatial Data Transfer Standard. A United States Federal standard designed to support the transfer of different types of geographic and cartographic spatial data. This standard specifies a structure and content for spatially referenced data in order to facilitate data transfer between dissimilar spatial database systems. Also known as Federal Information Processing Standard (FIPS) 173 (ESRI 1996).

SGML   Standard Generalised Mark-up Language. An ISO Standard defining the general framework for describing a document structure. This method of coding text is used for the storage of information on CD-ROM (Mobbs 1997).

Smallworld   A commercial object-oriented GIS package. Not widely used within archaeology.

SMR   Sites and Monuments Record.

SPANS   SPatial ANalysis System. A commercial GIS package capable of handling raster and vector data.

SPOT   Satellite Pour l'Observation de la Terre. A remote sensing satellite which has been developed by the French National Space Centre (CNES). The first SPOT (SPOT 1) was launched in February 1986, SPOT 2 was launched in 1988 (Walker 1993).

Stuffit   A file compression utility commonly used on Macintosh platforms.

TAR   A file compression utility commonly used on UNIX workstations.

TeX   A widely used document exchange format.

TIFF   Tagged Interchange File Format. An industry-standard raster data format. TIFF supports black-and-white, gray-scale, pseudocolor, and true-color images, all of which can be stored in a compressed or uncompressed format. TIFF is commonly used in desktop publishing and serves as an interface to numerous scanners and graphic arts packages (ESRI 1996).

TIN   Triangulated Irregular Network. A form of the tesseral model based on triangles. The vertices of the triangles form irregularly spaced nodes and unlike the DEM, the TIN allows dense information in complex areas, and sparse information in simpler or more homogeneous areas. The TIN data set includes topological relationships between points and their neighbouring triangles. Each sample point has an X,Y co-ordinate and a surface, or Z-Value. These points are connected by edges to form a set of non-overlapping triangles used to

	represent the surface. Tins are also called irregular triangular mesh or irregular triangular surface model (Walker 1993).
Topology	The study of relative relationships of geographic phenomena. When discussing digital data, topology generally refers to the relative relationships of points, lines, and polygons (after Walker 1993).
TSIP	Trimble Standard Interface Protocol. A proprietary satellite receiver output protocol.
UUENCODED	A format used to facilitate the transfer of binary files via e-mail.
UTM	Universal Transverse Mercator. This is a projection system based upon the Transverse Mercator projection. It is frequently used for the production of topographic maps and for georeferencing satellite images (Walker 1993).
VPF	Vector Product Format. This is a digital geographic vector-based format used by the US Defence Mapping Agency for the distribution of its vector data sets (ESRI 1996).
WGS 84	World Geodetic System 1984. This is a reference ellipsoid commonly used by satellite locational devices.
Word	A commonly used word-processing package.
WordPerfect	A commonly used word-processing package.
WWW	The World Wide Web facility of the Internet.

7.3 A QUICK REFERENCE GUIDE

7.3.1 Introduction

We appreciate that having carefully read the guide, many individuals and organisations will want quick and easy access to the recommendations held within, without having to search repeatedly through the full text. To this end we have produced this quick reference guide which contains all of the relevant information in a convenient bulleted form. Readers are reminded that more details and full contextual discussions are to be found within the main body of the guide. A comprehensive set of references are provided here to facilitate the easy location of this information when needed.

7.3.2 Spatial Data Models (section 3.2)

7.3.2.1 The Vector model (section 3.2.1)

The following information should always be recorded when assembling, compiling and utilising vector data:

- The data type, Point, Line or Area
- Type of topology which the file contains
- Details of any automatic vector processing applied to the theme
- State of the topology in the file
- Projection system
- Co-ordinate system

7.3.2.2 The Raster model (section 3.2.3)

The following information should always be recorded when assembling, compiling and utilising raster data:

- grid size (number of rows and columns)
- grid resolution
- georeferencing information, e.g. corner co-ordinates, source projection.

7.3.3 Attribute Data Models (section 3.10)

When attempting to structure and organise a flexible attribute database the following factors are of critical importance:
- Naming conventions
- Key fields
- Character field definitions
- Grid references
- Validation
- Numeric data
- Data entry control
- Confidence values
- Consistency
- Documentation
- Dates

7.3.4 Data capture techniques for map-based data (section 3.6.1.1)

7.3.4.1 Data capture using a Scanner (section 3.6.1.2)

- Details of the scanning device used, software driver and version
- All parameters chosen in the scanning process, such as the resolution setting of the device, the number of bits per pixel used
- Details of any pre-processing undertaken on the source mapsheet. This may include a range of options provided by the specific scanning software used
- Details of any post-processing undertaken on the data, such as noise reduction or sharpening with convolution filters, histogram equalisation, contrast adjustment

7.3.4.2 Data capture using a Digitiser (section 3.6.1.3)

- Detail of the digitising device used, software driver and version
- The precision, usually specified as a quoted resolution or as lpi
- Details of any automatic vector processing applied to the theme (such as snap-to-nearest-node)
- Details of control points used to manage conversion from digitiser to real-world planar co-ordinate systems
- Errors incurred in the above transformation process (e.g. quoted RMS)

7.3.4.3 Data Capture using a scanning-digitising hybrid (section 3.6.1.4)

In the case of using both a scanner and a digitiser to capture data, for example during 'heads-up digitising', the full information above for both the digitising and scanning procedures should be recorded.

7.3.5 Common Sources of Spatial Data (section 3.6)

- Maps and Plans
- Textual and numeric data
- Purchased or downloaded digital data
- Aerial photography
- Satellite and airborne remotely sensed images
- Terrestrial Survey data
- Satellite-based (GPS) data

7.3.6 Common Sources of Attribute Data (section 3.8)

Below are some likely sources of attribute data which you may come across, and wish to re-use:

- paper based card indexes
- archaeological site and survey archives (including paper based records, finds databases)
- qualitative report texts and articles published in journals (paper based or on the Internet)
- microfiche archives
- geophysical interpretation data derived from interpreted geophysics plots
- aerial photograph interpretations which may include morphological analysis attribute data and photo source information
- typological databases or artefact type series
- data generated at a regional level for integrated large scale historic landscape studies, such as the English Heritage Open Fields Project
- local level archaeological databases (e.g. Sites and Monuments Records or Urban Archaeological Databases where they are held separately from SMRs)
- local museum site and finds databases
- local Record Offices
- national archaeological databases (such as the various RCHM National Monument Records or English Heritage's database of Scheduled Ancient Monuments)
- Gardens Trust surveys
- historic buildings surveys and databases maintained by local authorities
- metadata relating to data sets

7.3.7 A more detailed look at Data Sources

7.3.7.1 Maps and plans (section 3.6.1)

In general the following information should always be recorded:

- Publisher **and** copyright owner
- The map medium

- Scale of source map
- Name of the map and the map series
- Claimed accuracy for any specific map components
- All details of the map projection and co-ordinate system employed

7.3.7.2 Textual and numeric data (section 3.6.2)

When integrating textual and numeric data the following information should be recorded:

- The data source
- The precision of the quoted co-ordinates
- Have the quoted locations been verified and how
- Projection system/co-ordinate origin
- If derived from a source map, record details of the map-base used
- If derived from a survey programme, record details of the survey procedure

7.3.7.3 Purchased or downloaded digital data (section 3.6.3)

Several standard formats and standards are of interest:

- British Standard 7567 (NTF: National Transfer Format), the format used by Ordnance Survey for the supply and transfer of digital products
- The recommendations of the National Geospatial Data Framework (NGDF)
- SDTS (Spatial Data Transfer Standard), a United States Federal Information Processing Standard (FIPS)
- DLG (Digital Line Graph) format, used by the USGS for supply of vector information
- DRG (Digital Raster Graphics), is the description that the USGS gives for the distribution of scanned map sheets
- DXF (Digital eXchange Format) format, commonly used for transferring drawings between CAD (Computer Aided Design) systems

7.3.7.4 Aerial photography (section 3.6.4)

To incorporate scanned and rectified aerial photographs into GIS databases the following information should be recorded:

- Full Photographic details
- Details of the scanning process
- Details of the rectification method(s) used
- The software employed including, where possible, specific parameters chosen
- Details regarding the ground control points (GCPs) used during the procedure
- Details of any post-processing undertaken on the data

7.3.7.5 Satellite and airborne remotely sensed images (section 3.6.5)

To incorporate remotely sensed data into GIS databases the following information should be recorded:

- Data source
- Date image was captured

- Data resolution
- Details of any post-processing undertaken on the data
- Details of the rectification method(s) used
- The software employed including, where possible, specific parameters chosen
- Details regarding the ground control points (GCPs) used during the procedure

7.3.7.6 Terrestrial survey (section 3.6.6.1)

When integrating data themes which are derived from survey data, the following should be recorded:

- The source and estimated error of survey base station co-ordinates
- Details of the survey, including date time and purpose
- Details of the thematic organisation of the survey
- Make and model of instrument used
- Type of survey (contour, feature etc.)
- Estimated error terms for the co-ordinate pairs and (if appropriate) the z-co-ordinate
- Georeferencing information, overall accuracy of the survey data

7.3.7.7 Satellite-based survey (GPS) (section 3.6.6.3)

In integrating GPS data the following information should be recorded:

- The method used to locate stations: C/A or P code pseudorange measurements, carrier phase measurements and whether a single measurement or averaging (include time period) was used
- The software used for any co-ordinate transformation and associated error estimate
- The satellites used in obtaining fix and observed GDOP (Geometric Dilution of Precision)
- The nature of any differential correction undertaken + error estimates
- The broadcast differential: name of the service provider and the name and location of base station
- The local base station: instrument details, location (including error estimate) of base station
- Post-processing: the software used and the source of correction data

7.3.8 Creating a GIS database (section 4.2)

When combining and integrating information from a variety of sources the following points should be kept in mind:

- All spatial data must be recorded in the same co-ordinate system. Data which are recorded to some other system must be transformed/projected to the required co-ordinate system.
- All spatial data should be to the same *spatial resolution*, or *scale*. It is not possible to get meaningful results from the combination of spatial data recorded to a *scale* of 1:250, as might be the case for an excavation site plan, with road alignments recorded to a scale of 1:250,000. Spatial data recorded to scales of greater than around 1:10000 involve considerable generalisation of alignments to avoid features conflicting. This is especially true of *paper* maps drawn to such scales.
- Non-spatial information to be combined, or integrated, must use the same field definitions,

encoding regimes, etc. Where different schemes are used it will be necessary to convert or translate the data to the required scheme.

7.3.9 Documenting the Data set (section 5.3)

Information about where the data you use are acquired from is one of the most important things you can record whilst constructing and using a GIS. The following comprises a non-exhaustive list of the information you might wish to record during your everyday creation, collection, and use of data:

- Computer hardware used
- Computer software used
- Date the data were captured/purchased/whatever
- Who did the work
- Data source ('bought from Ordnance Survey', etc.)
- Scale/resolution of data capture
- Scale/resolution at which data are currently stored
- Root Mean Squared error or other assessments of data quality
- Purpose of data set creation, where known
- Method of original data capture (Total Station Survey, etc.)
- Purpose for which you acquired the data (might differ from the previous information where the data were created by someone else for one purpose, and bought from them by you for another)
- Complete history of data ownership/rights

7.3.10 Supported GIS file deposit formats (section 6.3.3)

The ADS is able to accept most major file formats, though Arc/Info formats are preferred as this is the software which we actually use to manage and migrate GIS data sets. We can accept anything from a combination of DXF and DBF files to the following GIS formats:

- Arc/Info export
- Arc/Info ungen
- ArcView
- IDRISI (up to version 3)
- GRASS
- MIF/MID
- NTF
- SDTF
- MOSS
- VPF

Appendix 1: Data Sources

AERIAL PHOTOGRAPHY (VERTICAL ONLY) –
SOURCES ARRANGED ALPHABETICALLY BY COUNTRY

Austria

An archive of aerial photographs of archaeological sites is available from the Aerial Archive at the Institute for Prehistory and Protohistory at the University of Vienna (http://www.univie.ac.at/Luftbildarchiv/). Contact Michael Doneus at Michael.Doneus@univie.ac.at for more information.

Canada

Canadian Earth Observation Network (CEONET) offers a very large website with a variety of data archives and spatial databases (http://ceonet.ccrs.nrcan.gc.ca/). Aerial photography has been collected from a wide variety of agencies and companies in Canada and beyond. The website has a lot of information, but is somewhat unwieldy to navigate around and uses frames.

Slovenia

There is an extensive archive of aerial photographs covering the Republic of Slovenia. 1:20,000 vertical photographs can be obtained from the Ministry of the Environment and Planning (http://www.sigov.si/rgu/), and newer 1:17,500 (and larger) photographs are also available.

United Kingdom

National aerial photography coverage for the UK is available from the following organisations. If they do not have what you are looking for, they can probably point you in the right direction:

England	Northern Ireland	Scotland	Wales
English Heritage National Monuments Record Centre Kemble Drive Swindon SN2 2GZ	Ordnance Survey Colby House Stranmillis Court Belfast BT9 5BJ	RCAHMS John Sinclair House 16 Bernard Terrace Edinburgh EH8 9NX	Welsh Office Register of Air Photographs Planning 9 Room G-003, Crown Offices Cathays Park Cardiff CF1 3NQ and RCAHMW Plas Crug Aberystwyth SY23 1NJ

The Cambridge University Committee for Aerial Photography (CUCAP) has a collection of aerial photographs dating back to 1945. This collection includes aerial photographs taken by J. K. St. Joseph and his colleagues, and is one of the most important in the UK for archaeologists. CUCAP can be contacted at:

Cambridge University Collection of Air Photographs
The Mond Building
Free School Lane
Cambridge CB2 3RF
aerial-photography@cam.ac.uk
http://www.aerial.cam.ac.uk/

Another excellent resource for locating aerial photographs in the United Kingdom is a book entitled *Directory of Aerial Photographic Collections in the United Kingdom 1993* published by Aslib. The cost of this book is currently £22 (£18 for Aslib members), and copies can be ordered from:

Portland Press Ltd.
Commerce Way
Whitehall Industrial Estate
Colchester CO2 8HP UK
Telephone: +44 (0)1206 796351; Email: sales@portlandpress.co.uk

National Remote Sensing Centre (NRSC) located in the United Kingdom, distributes a variety of aerial photographs, and claims to have the largest colour A/P archive in the UK (http://www.nrsc.co.uk/).

Natural Environment Research Council (NERC) Scientific Services Data Centre – negatives of all NERC airborne photographs taken from 1982 to the present with associated maps, flight reports, and calibration details. Access depends on whether one is a NERC grant recipient, a member of the UK Higher Education community, or another user. See the NERC website for details (http://www.nerc.ac.uk/environmental-data).

United States

United States Geological Survey (USGS) has a wide variety of aerial photographs, but rectified and unrectified. The catalogue can be searched, and detailed ordering information is available online (http://mapping.usgs.gov/).

AERIAL PHOTOGRAPHY MAPPING AND ARCHAEOLOGICAL INTERPRETATION

In the above section, resources for finding aerial photographs were provided. In this section, sources for archaeological interpretation and mapping of the features on aerial photographs is covered.

Austria

Expert advice in interpreting archaeological features on aerial photographs is available from the Aerial Archive at the Institute for Prehistory and Protohistory at the University of Vienna. The website also provides helpful tips about rectifying and interpreting aerial photographs generally (http://www.univie.ac.at/Luftbildarchiv/).

Czech Republic

Please contact the following people for additional information:

Martin Gojda Institute of Archaeology Czech Academy of Sciences Letenska 4, Mala Strana 118 01 Prague 1 Czech Republic Telephone: +42 02 573 20 942 Fax: +42 2 539 361 gojda@arup.cas.cz	Jaromir Kovarnik South Moravian Museum Premyslovcu 6 6/669 02 Znojmo Czech Republic Telephone: +42 0624 224961 Fax: +42 0624 225210

Slovak Republic

For more information please contact Ivan Kuzma at:

Institute of Archaeology
Slovak Academy of Sciences
Akademicka 2
SK 949 21 Nitra
+421 87 357 38
+421 87 356 18 fax
Kuzma@savnr.savba.sk

Slovenia

Please contact the following people for archaeological aerial photographic interpretation and remote sensing:

Darja Grosman Department of Archaeology University of Ljubljana Zavetiska 5 1000 Ljubljana +386 61 262 782 phone +386 61 1233 082 fax darja.grosman@uni-lj.si	Zoran Stancic Centre for Scientific Research Slovenian Academy of Sciences and Arts Gosposka 13 1000 Ljubljana Slovenia +386 61 1256 068 phone +386 61 1255 253 fax zoran@zrc-sazu.si http://www.zrc-sazu.si/www/pic/

United Kingdom

The following organisations have collections that provide excellent starting places:

The Cambridge University Committee for Aerial Photography (CUCAP) has a collection of aerial photographs dating back to 1945. This collection includes aerial photographs taken by J. K. St. Joseph and his colleagues, and is one of the most important in the UK for archaeologists. CUCAP can be contacted at:

Cambridge University Collection of Air Photographs
The Mond Building
Free School Lane
Cambridge CB2 3RF
aerial-photography@cam.ac.uk
http://www.aerial.cam.ac.uk/

England	Northern Ireland	Scotland	Wales
English Heritage NMRC Kemble Drive Swindon SN2 2GZ http://www.english-heritage.org.uk/	Environment & Heritage Service Monuments & Buildings Record 5–33 Hill Street Belfast BT1 2LA http://145.229.156.11:8080 /index.htm	RCAHMS John Sinclair House 16 Bernard Terrace Edinburgh EH8 9NX http://www.rcahms.gov.uk/	RCAHMW National Monuments Record Crown Buildings Plas Crug Aberystwyth SY23 2HP http://rcahmw.gov.uk/

English Heritage NMR and the two Royal Commissions for Scotland and Wales have expert staff trained in air photo interpretation to provide advice for customers.There are also private consulting firms which provide aerial photographic interpretation services especially for archaeologists. In the United Kingdom these include:

Air Photo Services Contact: Rog Palmer 21 Gunhild Way Cambridge CB1 4QZ	Air Photo Services Ltd Contact: Chris Cox 7 Edward Street Cambridge CB1 2LJ

Help and advice is also available from the Aerial Archaeology Research Group which can be contacted via:

Cathy Stoertz
Chair
Aerial Archaeology Research Group
English Heritage
c/o Aerial Survey
Kemble Drive
Swindon
SN2 2GZ

Remote Sensing Applications in Archaeology is a website that provides a list of annotated links to data suppliers. It also provides descriptions of archaeological projects in which aerial photography is actually being used so you can see what others projects are doing (http://eleftheria.stcloudstate.edu/arsc/).

GIS DATA

Bartholomew Digital Map Data can be ordered over the World Wide Web. Special discounted prices on this data are available to UK Higher Education institutions via CHEST. Contact your campus CHEST advisor for details. Data available includes 1:5,000 coverage for London and 1:250,000 coverage for the rest of Great Britain (http://www.mimas.ac.uk/maps/barts/barts1.html).

Canadian Earth Observation Network (CEONET) offers a very large website with a variety of data archives and spatial databases. Resources of interest to GIS users include the National Atlas of Canada Digital Data and the National Topographic Database for Canada. The website has a lot of information, but is somewhat unwieldy to navigate around and uses frames (http://ceonet.ccrs.nrcan.gc.ca/).

The Centre for Advanced Spatial Technologies provides a useful webpage called Starting the Hunt: Guide to On-line & Mostly Free U.S. Geospatial and Attribute Data compiled by Stephan Pollard. Great for research in the United States (http://www.cast.uark.edu/local/hunt/index.html).

Environmental Systems Research Institute (ESRI), the company that produces Arc/Info and Arcview, has a variety of GIS data sets available, some of these are free GIS data sets (http://www.esri.com/).

Macaulay Land Use Research Institute – Scotland provides a variety of land cover and soil data sets for Scotland. They can be contacted at:

Macaulay Land Use Research Institute
Craigiebuckler
Aberdeen AB9 2QJ
Tel: (01224) 311 556

MIDAS (now known as MIMAS) is a service at the University of Manchester which provides access to Bartholomew digital map data and 1981 and 1991 Census of Population digital boundary data for the UK Higher Education sector. Site licenses are purchased by UK universities on behalf of their employees and students. Contact your campus representative for details, or email info@mimas.ac.uk for assistance (http://www.mimas.ac.uk).

NOAA Environmental Services Data Directory is a one-stop shop to find out about many data sets, both digital and non-digital, that relate to the study of the past and present environment (http://www.esdim.noaa.gov/NOAA-catalog/).

The Ordnance Survey is the UK mapping agency, and thus offers a wide variety of data useful for GIS analysis. A product index provides descriptions of the data sets. These include raster and vector data at a variety of scales up to 1:1,250, DEMs, boundary data, and address point data (http://www.ordsvy.gov.uk/).

Rutgers University and USA-CERL market CD-Roms containing global environmental data. These CD-Roms are specifically designed for use with Arc/Info, Arcview, ERDAS IMAGINE, and GRASS and contain such interesting coverages as "pigs per square kilometer" and "camels per square kilometer" as well as the more run-of-the-mill soil, political boundaries, rivers, etc. This is small-scale data so it's appropriate for global modelling rather than site-specific archaeological overlays. Some sample images are available (no camels, though). Cost for the Arc/Info and Arcview CD-Rom is $375 plus $10 postage and packing (http://deathstar.rutgers.edu/global.html).

NASA's Global Change Master Directory is a good web site for those who know precisely the data they want. The interface to this large database is primarily via user-defined searches rather than pull-down menus (http://gcmd.gsfc.nasa.gov/).

Spatial Information Enquiry Service (SINES) is a web-based query system that allows you to search for any data available from the Inter-departmental Group on Geographic Information (IGGI). This includes data from the British Geological Survey, Countryside Commission, Department of Environment, English Nature, Forestry Commission, Ministry of Defense, National Rivers Authority, Ordnance Survey, Scottish Office, and the Welsh Office (http://www.ordsvy.gov.uk/sines-bin/SearchCGI).

United States Geological Survey (USGS) distributes a wide variety of GIS data sets including Digital Line Graphs (DLG), Digital Raster Graphics (DRG), Digital Elevation Models (DEM), Land Use and Land Cover files (LULC), and regional U.S. data sets. The catalogue can be searched, and detailed ordering information is available online (http://mapping.usgs.gov/).

REMOTELY SENSED DATA

Canada Centre for Remote Sensing (CCRS) – markets a wide variety of satellite products that cover Canada. This includes SPOT, Landsat Thematic Mapper, Landsat Multispectral Scanner, AVHRR, Seasat, ERS-1, Jers-1, Radarsat, and MOS MESSR data. A searchable catalogue and quicklook images are available (http://www.ccrs.emr.ca/).

Canadian Earth Observation Network (CEONET) offers a very large website with a variety of data archives and spatial databases. Most of the satellite imagery comes from CCRS (see above), and its website is easier to search. The CEONET website has much more information, but is somewhat unwieldy to navigate around and uses frames (http://ceonet.ccrs.nrcan.gc.ca/).

Committee on Earth Observation Satellites (CEOS) International Directory Network (CEOS IDN) – This is a (slightly slow) website that helpfully provides a central point for information about satellite imagery. Data is clustered into four nodes: Africa, America, Asia, and Europe. This site is a good source of data from Argentina, Australia, Brazil, Canada, France, Germany, Italy, Japan, the Netherlands, and the United States. Future links will add data from China and Russia (http://gcmd.gsfc.nasa.gov/ceosidn/).

Eurimage distributes multi-mission satellite data, and specialises in coverages for Europe, North Africa, and the Middle East. Satellite products include Landsat TM, KVR-1000, RESURS-01, MK-4, ERS, AVHRR, JERS-1, and KFA 1000. Some quicklooks and metadata are available at no charge, but access to their full catalogue requires payment of a subscription. The Eurimage website provides a useful page of service providers which includes links to FAQ pages (http://www.eurimage.it/).

European Space Agency's earthnet*online* offers quick searching of a variety of satellite products including Landsat, ERS, and JERS (http://gds.esrin.esa.it/).

MIDAS (now known as MIMAS) is a service at the University of Manchester which provides access to UK coverages of Landsat and SPOT data for the UK Higher Education sector. Site licenses are purchased by UK universities on behalf of their employees and students. Contact your campus representative for details, or email info@mimas.ac.uk for assistance (http://www.mimas.ac.uk/).

NASA's Global Change Master Directory is a good web site for those who know precisely the data they want. The interface to this large database is primarily via user-defined searches rather than pull-down menus. Expect to find information about commonly-used remotely sensed data (e.g. Landsat, SPOT, AVHRR) (http://gcmd.gsfc.nasa.gov/).

National Remote Sensing Center (NRSC) located in the United Kingdom, provides a processing and archiving facility for ERS satellite products. NRSC is also licensed to sell a wide variety of other satellite products including Landsat MSS and TM, SPOT, IRS, JERS, SAR, and various products from Russian Satellites (i.e. KFA1000, KFA3000, MK4, Kate200, KVR1000, TK350, and MSU-SK.) Their website has a nice user interface, and clearly describes each of the sensors onboard the various satellites and the types of images that can be derived from each (http://www.nrsc.co.uk/).

Natural Environment Research Council (NERC) Scientific Services Data Centre – an extensive archive and catalogue of satellite and airborne data including 2000 tapes of non-NERC data (e.g. Landsat Thematic Mapper, Landsat Multispectral Scanner, and some SPOT scenes) plus NERC data from 1982 to the present (e.g. Airborne Thematic Mapper and some hyperspectral sensor data). Access depends on whether one is a NERC grant recipient, a member of the UK Higher Education community, or another user. See the NERC website for details (http://www.nerc.ac.uk/nss/).

Remote Sensing Applications in Archaeology is a website that does not actually provide data, but does provide a list of annotated links to data suppliers. It also provides descriptions of archaeological projects in which satellite imagery is actually being used so you can see what data others are using (http://eleftheria.stcloud.msus.edu/rsaa/).

SPOT – satellite systems designed by the Centre National d'Etudes Spatiales in France. Metadata for more than 4,000,000 SPOT images are available through the DALI catalogue and DIVA metadata handling tools. This catalogue is updated daily. Scenes can be searched over the Internet via a limited range of variables, but complete metadata records are available to registered users (current cost for this service is FFr 500 or roughly US$100 for one year). A CD-Rom version of the catalogue, containing 60% of SPOT images collected since 1986, is available for the cost of postage, packing, and media. Technical information about the satellites, their orbit, and the SPOT images is available (http://www.spot.com/).

United States Geological Survey (USGS) satellite imagery includes NASA photographs, AVHRR images, Landsat Thematic Mapper data, and Landsat Multispectral Sensor data. The catalogue can be searched, and detailed ordering information is available online (http://mapping.usgs.gov/).

Appendix 2: Standards in Archaeology

DATA DOCUMENTATION AND CONTENT STANDARDS

Formal Standards

Archiving Aerial Photography and Remote Sensing Data: a Guide to Good Practice (Bewly et. al 1999) – Written by a working party of archaeological aerial photography and remote sensing specialists, and widely peer reviewed, this guide deals with all aspects of creating, documenting, and archiving images and interpretations. The creation of Dublin Core–style metadata records for these data is also described, and examples are provided (http://ads.ahds.ac.uk/project/goodguides/apandrs).

Art and Architecture Thesaurus – The Getty Information Institute produced this useful thesaurus in 1990 and has recently made it searchable over the World Wide Web. It's an invaluable tool for standardised description of material culture, architecture, and art in the Western World from prehistory to the present. Vocabulary is controlled through a hierarchical structure of broader/narrower terms, synonym control, and other helpful tools (http://www.gii.getty.edu/aat_browser/).

British Archaeological Thesaurus – Written by Cherry Lavell and published by the Council for British Archaeology in 1989, this was the earliest thesaurus for British archaeology. It remains a useful companion to back issues of the *British and Irish Archaeological Bibliography* (and its predecessors, *British Archaeological Bibliography* and *British Archaeological Abstracts*), but is now out-of-print.

British Museum *Materials Thesaurus* – Released and published in 1997 at the same time as the MDA Archaeological Object Thesaurus, this thesaurus was edited by Tanya Szrajber and compiled by the museum Collections Data Management Section. This thesaurus was a response to their need to document collections from any time period around the world, and it provides a good resource for terminology control of organic, inorganic, and processed materials (http://www.open.gov.uk/mdocassn/bmmat/matintro.htm).

Data Standards and Guidance for Digital Data Transfer to the Northamptonshire SMR Developed by Northamptonshire Heritage, this document covers all aspects of archaeological practice in Northamptonshire from project design through data collection and digital archiving. All contractors operating in Northamptonshire are expected to comply with recommendations in this document.

Elib Standards Guidelines (Version 2). This covers, concisely, a wide range of electronic format and interchange standards, and includes references to more detailed reading (http://www.ukoln.ac.uk/services/elib/papers/other/standards/.

Guide to the Description of Architectural Drawings – Published in 1997, this guide provides a general introduction to the principles of documenting architectural materials with recommendations for both digital and manual systems. Developed by the Getty Information Institute and partners (http://www.gii.getty.edu/fda/index.html).

International Guidelines for Museum Object Information: CIDOC Information Categories CIDOC, the International Documentation Committee of the International Council of Museums has developed guidelines about what information should be recorded for museum objects, how it should be recorded, and the terminology with which this information should be recorded. This standard is particularly useful for archaeologists working in a museums setting (http://www.cidoc.icom.org/guide0.htm).

Lincolnshire Archaeological Handbook – Published by Lincolnshire County Council, this handbook establishes criteria for the conduct of archaeological projects. Covers such topics as project management, archaeological methods, recording systems, archive preparation, dissemination, and all other phases of archaeological research. This resource is vital for any unit or university research project undertaking work in Lincolnshire (http://www.lincscc.u-net.com/archindx.htm).

Management of Archaeological Projects (MAP2) – Developed by English Heritage as a guide to the management of all phases of archaeological projects. Includes guidelines for planning, fieldwork, assessment of potential, analysis, report preparation, and archiving (http://www.eng-h.gov.uk/guidance/map2/.

MIDAS: A Manual and Data Standard for Monument Inventories – Developed by RCHME in 1998 to assist in the creation of monument inventories such as Sites and Monuments Records (SMRs). Includes recommendations for terminology control for archaeological evidence in England, and is an indispensable resource for archaeologists working in England. RCHME hope to make a usable flat-file version available on their website soon.

MDA Archaeological Objects Thesaurus – The Museums Documentation Association has released this thesaurus of object and artefact names in conjunction with the Royal Commission on the Historical Monuments of England and English Heritage (mda 1997). The goal of this thesaurus is to encourage access to, and reuse of, collections, archives and record systems, and to facilitate cooperation and data exchange between all individuals and institutions involved in the retrieval, research and curation of archaeological objects.

NPPG5 - Archaeology and Planning. This National planning policy guideline has been produced by The Scottish Office and sets out the Government's planning policy on how archaeological remains and discoveries should be handled under the development plan and development control systems, including the weight to be given to them in planning decisions and the use of planning conditions (http://www.scotland.gov.uk/library/nppg/nppg5con.htm).

Rules for the Construction of Personal, Place and Corporate Names – The National Council on Archives' 1997 guide to the recording of name information in archives. These rules include guidance on the use of non-current place names, and other issues of relevance to the archaeological community. Available from http://www.hmc.gov.uk/nca/title.htm.

TGN – The Thesaurus of Geographic Names is a project of the Getty Information Institute, aiming to create a powerful resource holding information on names of inhabited places, regions and geographic features both now and in the past. Importantly, TGN is holding these names within a hierarchy, such that it may be determined that a town lies *within* a country, that country *within* a continent, etc (http://www.gii.getty.edu/tgn_browser).

Thesaurus of Monument Types: A Data Standard for Use in Archaeological and Architectural Records – The second edition produced by the Royal Commission on the Historical Monuments of England (RCHME) in 1998, this is a standard for use with both archaeological and architectural information. This thesaurus is actively updated by the Data Standards Unit at the RCHME. The purpose of this thesaurus is to standardise the terms used to describe archaeological sites or standing buildings by, for example, listing terms hierarchically and relating the levels of this hierarchy to one another or indicating preferred terms in the case of synonyms. For example, the hierarchical structure means that RELIGIOUS monuments include the subset of MONASTERYs and that there is a further sub–division into BENEDICTINE MONASTERY or CISTERCIAN MONASTERY depending on the particular monument being described. Synonyms are dealt with by pointing the user to a preferred term (e.g. for 'tribunal' use COURT HOUSE). This is one of the most widely used documentation standards in UK archaeology, and is a useful source of subject terms.

Draft Standards

Coin Recording Standard – Currently packaged as an appendix to the Portable Antiquities Recording Scheme database user manual, this standard for recording coins is helpful for amateur and professional archaeologists.

International Core Data Standard for Archaeological Sites and Monuments (Draft) – Produced in 1995 by CIDOC, the International Documentation Committee of the International Council of Museums, this document guides the user in documenting archaeological sites and monuments. The goal of this standard is to facilitate international exchange of information by encouraging standardised approaches to database structure. Useful information about naming, describing, cross–referencing, and spatially referencing sites and monuments is provided. Working examples from Denmark, England, France, and the Netherlands are provided. Contributors come from these countries and Albania, Canada, Poland, Romania, Russia, and the United States (http://www.natmus.min.dk/cidoc/archsite/coredata/arch1.htm).

National Heritage Reference Dataset – Currently being developed by FISHEN – the Forum on Information Standards in Heritage (England) – this resource is intended for use with MIDAS. A variety of termlists and scope notes have been developed for high-level terminology control for things such as scientific dating techniques used in archaeology (http://www.mda.org.uk/fishen/).

Planning and the Historic Environment (DRAFT). A draft copy of national planning policy guideline produced by The Scottish Office which sets out the Government's policies for the planning of development in historic environments with a view to their protection, conservation and enhancement. It has been prepared on the basis of the existing statutory framework for planning and reinforces current legislation relating to listed buildings and conservation areas for which the Secretary of State's executive agency, Historic Scotland, is primarily responsible (http://www.scotland.gov.uk/library/nppg/pathe-00.htm).

SMR'97 (Draft) Currently being developed by the Association of Local Government Archaeological Officers (ALGAO), English Heritage, and the Royal Commission on the Historical Monuments of England this data standard is designed to unify the structure of Sites and Monuments Records databases throughout England.

Strategy for Recording and Preserving the Archaeology of Wales Produced in 1998 by Cadw and the Royal Commission on the Ancient and Historical Monuments of Wales and endorsed by the Ancient Monuments Board for Wales, this draft sets the agenda for recording, protection, and management of the archaeological resource in Wales.

Informal Standards and Helpful Resources

INFO2000 – The European Commission's INFO2000 website contains extensive list to all kinds of standards. This is a very, very useful resource (http://www2.echo.lu/oii/en/oiistand.html).

JISC/TLTP Copyright Guidelines. This document, available as a pdf file, is targeted at the HE audience and covers a wide range of copyright issues in electronic media. A very useful document for reference (http://www.ukoln.ac.uk/services/elib/papers/others/jisc-tltp/jisc.pdf).

*word*HOARD – The mda wordHOARD site is a good resource. It provides connections to all manner of standards – common and not – including details for such things as SPECIES 2000, the Smithsonian Institution's World List of Insect Families, and the Vascular Plant Familes and Genera databases from the Royal Botanic Gardens (http://www.open.gov.uk/mdocassn/wrdhrd1.htm).

SPATIAL DATA STANDARDS

Standards geared specifically towards the handling of map–like information are as numerous and diverse as those for other areas and, given the ADS responsibility for geospatial data standards across the whole Arts & Humanities Data Service, they are addressed separately here.

BS 7666 – British Standard 7666 specifies the manner in which address information should be specified, and is likely to prove extremely important within Local Government and the Utilities. Archaeologically, it may prove most useful in the consistent provision of address information for Listed Buildings, etc.

CEN prEN 287009 – This draft standard from the European Standards Organisation (Comité Européen de Normalisation) Technical Committee TC/287 addresses standards related to geospatial data, including the definition of a reference model, geometry guidelines, data description structures and data transfer issues.

FGDC – the United States' Federal Geographic Data Committee's Content Standard for Digital Geospatial Metadata is, perhaps, the best known and established of geospatial data standards, the current version having been released in 1998. This standard underpins much of the US Federal Government's work with geospatial data, and is also used by other collectors of spatial data, although there do not appear to be examples of its use within the UK archaeological community (http://www.fgdc.gov/metadata/contstan.html).

INFO2000 GIS page – The European Commission's INFO2000 programme has a wonderful website listing all varieties of standards. The GIS page provides a good set of links to spatial data standards (http://www2.echo.lu/oii/en/gis.html).

ISO 15046 – This draft standard from the International Standards Organisation Technical Committee TC/211 addresses geographic information and geomatics. The draft standard appears modelled upon current FGDC practice, and offers powerful options for extensibility and modification within the wider standards framework.

National Geospatial Data Framework – This important co–operative initiative is aiming to provide effective means of access to geospatial data collected and held by government and the public and private sectors following a model potentially similar to that of the AHDS. NGDF is addressing issues such as metadata standardisation and is looking to greatly increase the market for existing and new data (http://www.ngdf.org.uk/).

OGIS – The Open Geodata Interoperability Specification (OGIS) is an initiative of the vendor–led Open GIS Consortium (OGC). OGC is looking to increase the ease with which geospatial information may be passed between products, and OGIS is one important aspect of this work (http://www.opengis.org/).

SABE – The Seamless Administrative Boundaries of Europe (SABE) project is an initiative of the MEGRIN Project Team, looking to establish a single uniform set of administrative boundaries for the European Union. If successful, this could form an important underpinning to any spatially driven search interface (http://www.ign.fr/megrin/sabe/sabe.html).

UK Standard Geographic Base – The UK Standard Geographic Base is a proposal from the Office for National Statistics to develop a single framework by which spatial units within the UK may be uniformly described. This proposal is currently at the stage of having a sound business case developed before work proceeds.